Biotech's

DICTIONARY *Of* Agriculture Engineering

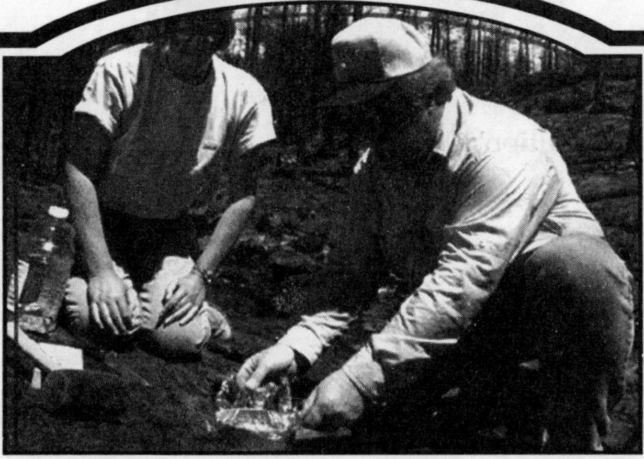

Compiled & Edited by
Dinesh Arora

BIOTECH
2005
BIOTECH BOOKS
DELHI - 110035

Biotech's

DICTIONARY *Of* Agriculture Engineering

Compiled & Edited by
Dinesh Arora

First Indian Edition 2005

ISBN 81-7622-134-1

Published by : BIOTECH BOOKS
1123/74, Tri Nagar,
Delhi - 110 035
Phone: 27382765
e-mail: biotechbooks@yahoo.co.in

Showroom : 4762-63/23, Ansari Road, Darya Ganj,
New Delhi - 110 002
Phone: 23245578, 23244987

Printed at : Chawla Offset Printers
New Delhi - 110 052

PRINTED IN INDIA

PREFACE

Agriculture Engineering and Technology is certainly one of the most rapidly advancing of the agricultural sciences, and probably no other field has stimulated so many divers disciplines. For example fields like agronomy, hydrology, instrumentation, nuclear agriculture, sewerage, waste water use, plant physiology, soil science, agricultural economics, meteorology, soil conservation, irrigation, drainage, farm machinery, post-harvest technology, and ecology are just few of them. Such growth is, of course, accompanied by a proliferation in terminology. Further many new terms are coined which are not found in standard dictionaries, and thus create a problem for students. This becomes the prime reason for carrying out specialised dictionaries like – Dictionary of Agriculture Engineering. In fact a dictionary of such a kind has an ocean of information which tries to satisfy us for our unending thirst for knowledge.

It has been our duty to condense the definitions, often a difficult matter when a longer explanation would have been easier to draw up. We trust that we have in each case succeeded in setting out the main or central meaning. It is hoped that the students, teachers, and researchers and all those interested in agriculture in general and

Agriculture Engineering in particular will find many of the terms commonly encountered in the Agriculture Engineering and allied literature defined in this book, and that it will encourage the development of a vocabulary which will aid a student in pursuing further studies in this fascinating field.

- **3, 4, 5 triangle corner method**
a technique used for establishing a right angle based on the Pythagorean theorem.

- **3-way switch**
an electrical switch with three terminals. It does not have marked on or off positions and is always used in pairs, allowing a load to be controlled from two locations.

- **4-way switch**
an electrical switch with four terminals. It does not have marked on or off positions and is always used with a pair of 3-way switches to control a load from three or more locations.

- **A**
soil loss per unit area in universal soil loss prediction equation of Wischmeier and Smith.

- **A frame ridger**
an A shaped wooden implement for making ridges and bunds.

- **A horizon**
1. the uppermost alluvial layers of a soil profile.
2. the surface horizon of a mineral soil having maximum biological activity or eluviation or both.

- **A(B)C soil**
a soil with three horizons in which the upper, A horizon is differentiated from B primarily by weathering.

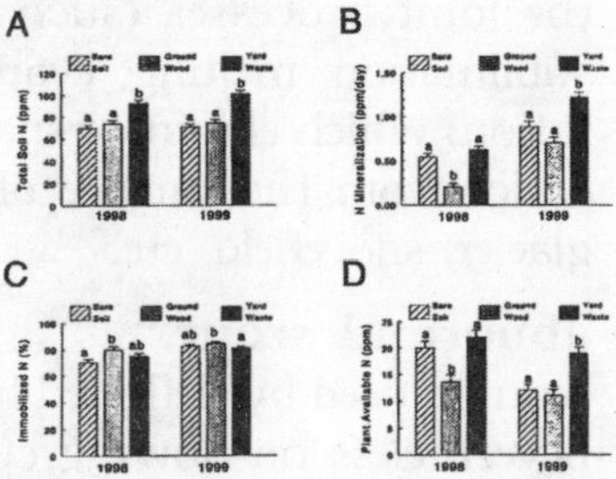

- **AB horizon**
a horizon in which leaching and accumulation balance out.

- **abatement**
the method of reducing the degree or intensity of pollution; also the use of such a method.

▪ **ABC soil**
soil with clearly developed profile, including, A, B and C master horizons.

▪ **ABD soil**
a soil in which the thin parent rock has been entirely transformed.

▪ **abiotic**
description of the non-living part of an ecosystem.

▪ **abiotic factors**
physical, chemical and other non-living components of the environment such as light, temperature, moisture, nutrients and other edaphic factors.

▪ **ablation**
the joint processes (such as sublimation, melting, evaporation) which eliminate snow or ice from the surface of a glacier, snowfield, etc.

▪ **abnormal scour**
scour caused by deflected but nevertheless on-flowing currents, as by the river running tangentially to a curve and forming an eddy where river leaves the curve of the protection bund.

▪ **abnormal soil**
an undeveloped soil in accordance with known laws in the climate considered.

▪ **abrasion**
mutual rubbing of particles of sediment in flowing water as well as in the bed of a channel as the result of which these particles tend to decrease in size.

▪ **abrasive process**
cutting which is performed by stones, sand, emery glass, silicon carbide, chilled globules of iron, and in some cases by soft friable iron alone.

▪ **ABS**
alkyl benzene sulphonate.

▪ **absolute artesian well**
an artesian well which is isolated by sealing from a permeable layer, if it happens to traverse it, overlying the upper confining bed of the artesian ground water.

- **absolute drought**
  a period of at least fifteen consecutive days, none of which has received as much as 0.25 millimetres of rain.

- **absolute water requirement**
  quantity of water in acre-inches per crop-season absorbed by the crop together with the evaporation from the crop-producing land.

- **absorbed water**
  water held mechanically in a soil mass and having physical properties not substantially different from ordinary water at the same temperature and pressure.

- **absorbing well**
  a shaft or well driven through an impermeable stratum to allow surface water to drain through to a permeable one.

- **absorption dynamometer**
  a device in which the energy measured is absorbed as frictional or electrical resistance and dissipated as heat.

- **absorption dynamometer**
  a tool for measuring power output.

- **absorption hygrometer**
  a hygrometer for determining relative humidity of the atmosphere.

- **absorption losses**
losses of water from a canal or reservoir by capillary action and percolation, and in the case of a canal, during the process of priming.

- **absorption pit**
a hole in the ground for disposal of rainwater, sillage or treated sewage effluent.

- **absorption ratio**
the ratio of the final internal concentration of a particular ion in plant tissue (*i*) to the final concentration *(e)*, That is i / e.

- **absorption terrace**
a terrace in the form of a ridge whose purpose is to slow down the flow of runoff water and retain it until it can be absorbed by the soil.

- **absorption test**
a physical test made upon porous material such as rock, concrete, clay, tile, etc., to determine the quantity of water, which it will absorb.

- **absorptive agent**
any material or chemical compound which possesses the ability or tendency to absorb water or other liquid products.

- **absorptive capacity**
the total capacity of a material to absorb a liquid, such as soil to absorb water.

- **absorptive outlet**
it is an outlet channel in which contour furrows, ridges, ditches, or other such means are used to collect and hold runoff until it is absorbed by the soil.

- **absorptive terrace**
a terrace of variable cross-section constructed by excavating on both sides of the ridge, formed and designed primarily for the retention and spreading of moisture over the widest possible area.

- **abstracting**
the process of collecting together prices of work of same sort having the same unit of price and measurement.

- **abutment pier**
a heavier pier, usually every fourth or fifth pier, built in an arched bridge (with series of arches) designed to withstand

heavy unbalanced inclined thrust from the superstructure (if some of the arches give way).

- **Abyssinian pump**
a pump having a well tube attached to the suction tube for use in an Abyssinian well.

- **Abyssinian well**
a tube with holes above the pointed end, driven into strata of moderate hardness to obtain a supply of water.

- **accelerated erosion**
erosion which takes place more rapidly or at a greater rate than the geologic norms.

- **access eye**
an access hole at a bend in a drain pipe, ordinarily covered by a metal plate wedged or bolted over it which enables the pipe to be rodded.

- **access gallery**
a horizontal gallery connecting the gallery system in the dam with the downstream face or features outside the dam, such as power house or gate house.

- **accessory system**
engine systems without which engine operations can be easily carried.

- **accommodation**
the capacity of a plant to adapt itself to changes in the environments provided that the changes occur gradually.

- **accretion**
any process of accumulation by flowing water whether of silt, sand, pebbles, etc. The difference between, accretion and alluviation is that the latter is due to retardation of flow whereas the former may be due to any cause and in fact includes alluviation.

- **accrued depreciation**
the accumulated depreciation which has taken place in a given structure, machine, etc., from the date when it was put in service or commenced to be used, to a given date.

- **accumulated**
it is the total quantity of water that enters the soil in a given time.

- **accumulated temperature**

a measure of the variation of temperature from a given norm for a particular place over a specified period, measured in degree-days. This is most commonly used for determining whether an area is suitable for the cultivation of particular crops.

- **accumulation curve**

it is a curve obtained by plotting settling time on the abscissa and the weight of the deposited particles on the ordinate. It represents the total amount of accumulated material at various time intervals.

- **accumulation outlet**

a means of disposing of the discharge from terraces and consists of ponds or dugouts which can be used to provide water for stock, recreation or wild life.

- **accumulative escaping power**

progressive totals at end of each reach of the escaping power from the first reach downstream. The design section must make provision for this discharge in addition to calculated theoretical discharge.

- **acetometer**

a device, which is used to estimate the amount of acetic acid in any solution, especially in vinegar.

- **acetylene welder**

a machine used for fusion welding or heating deriving its heat from the combustion of oxygen and acetylene stored in separate cylinders.

- **achromobacterium**

a genus of bacteria commonly found in activated sludge, trickling filters, de-nitrification processes and anaerobic sludge digestion.

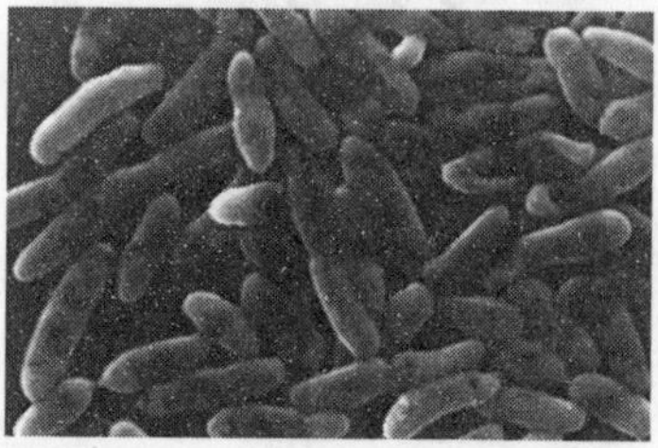

- **acid fermentation**

one stage of anaerobic decomposition of sewage sludge, resulting in the biodegradation of complex organic compounds to simpler organic acids, as in anaerobic sludge digestion.

- **acid formers**

a group of facultative and obligate anaerobes, capable of hydrolysis, which ferment the complex organic compounds in sewage to organic acids, including acetic acid (CH_3COCH), and propionic acid (CH_SCH_2COOH). For effective anaerobic sludge digestion they should be in equilibrium with the methanogenic bacteria. The main acid formers are the obligate anaerobes, which are 10-100 times more numerous in digesting sludge than the facultative anaerobes.

- **acid hydrolysis**

a method of converting the cellulose in municipal refuse into a non-polluting fuel-ethanol (ethyl alcohol).

- **acid soil**

a soil which is deficient in available bases, particularly calcium, and which gives an acid reaction when tested by standard methods.

- **acidification**

the process of removing incrustation of lime in strainers or screens by pouring sulphuric or hydrochloric acid into the well, allowing it to remain several hours and then surging to

dislodge the loosened material. It is also used for opening fissures in chalks and limestones.

- **acidulous water**

mineral water charged naturally with carbon dioxide and water containing sulphur compounds, especially sulphates.

- **acme harrow**

a harrow constructed with long, flat, twisted, knife blades that project rearward. It is used to cut clods and to stir the soil.

- **acme standard 29° thread**

a form of thread used to transmit power, of sides of which form an included angle of 29°.

Also used in sanitary pipe fittings.

- **acoustic sounding**
the indirect evaluation of water depth, using the principle of measuring the length of time necessary for a sound wave to travel to the bottom, reflect and travel back to the water surface.

- **acquired profile characteristics**
are characteristics more local in nature and are mainly due to the process of synthesis, eluviation and alluviation.

- **across the slope system**
a system of tile drains consisting of running the lines of the tiles across rather than down the slope.

- **activated alumina**
alumina ($A1_2O_3$) that has been treated to increase its surface area and thus to enhance its adsorptive properties. It can remove phosphate and arsenic compounds from liquid wastes and can dry gases.

- **activation**
the process of causing a substance to become artificially radioactive by subjecting it to bombardment by neutrons or other particles.

- **activation analysis**
an analytical procedure permitting the detection and measurement of trace quantities of elements following their exposure flux of neutrons.

- **activation energy**
the energy required before the reaction of concern will proceed. Often heating will supply the needed energy. Enzymes or other catalysts lower the activation energy of a reaction.

- **activation rancidity**
the process of causing milk to become rancid by adding warm milk to cold, too much stirring, or in some other way causing the milk to foam.

- **active capacity**
storage available for project purposes, usually the storage between the lowest point of release and the highest controlled water surface (static full pool level).

- **active earth pressure**
the minimum value of earth pressure. This condition exists

when a soil mass is permitted to yield sufficiently to cause its internal shearing resistance along a potential failure surface to be completely mobilised.

- **active element**
a nutrient in the soil that can be used by plants.

- **active equilibrium**
plastic equilibrium obtained by an expansion of a mass.

- **active humus**
humus which can become mineralised.

- **Active Ingredient (A.I.)**
the active component of a formulated product such as fungicide, e.g. Ridomil 25 wp and 75 per cent inert material. The ridomil has 25 per cent A.I.

- **active layer**
the thickness of soil frozen in winter, that thaws out in summer.

- **active water**
water having corrosive qualities.

- **active zone of well**
the zone from where the well is fed.

- **actual crop evapotranspiration**
rate of evapotranspiration equal to or smaller than PET as affected by the level of available soil water, salinity, field size, or other causes; mm/day.

- **Actual Evapotranspiration (AE)**
the total amount of water actually lost by a crop-soil complex through evaporation and transpiration.

- **actual farm yield**
the yield of a crop/variety on the farmer's field under optimum management practices.

- **actual water income**
water received from precipitation as well as that entering the area as underground flow available for transpiration within the area.

- **acute toxicity**
a condition of polluted air or water that kills quickly. It can be measured by a median tolerance limit.

- **adaptability of crop**
it is the crops ability to survive and reproduce in diverse environments.

- **adaptation**
a change in structure or habit or an organisms that produces better adjustment to the environment; a genetically determined characteristics that

enhances the ability of an organism to cope with its environment.

- **adapter**
a device which is used as a fitting to connect two variable products such as pipes of different sizes. An added device, which allows using a product for purposes other than that originally intended.

- **added water**
water which is combined legally or illegally with any product increasing its weight and volume. It is legally added in controlled amounts to sausage and canned fruit and vegetables. It is sometimes added to other products.

- **addendum**
an addition, that portion of a gear which projects beyond the pitch circle. An addition or supplement as of a book. Part of a tooth of a gear wheel located between the pitch line and the extreme point of a tooth.

- **additional outlet**
an outlet constructed at the expense of the landholder(s) to provide a further point of supply over and above that provided by the regular outlet.

- **adherence of a soil**
the capacity of a soil to adhere to foreign objects, particularly to the blades of farm implements.

- **adhesive water**
water forming a film around soil particles over adsorption water and held by the forces of molecular attraction, but with less strength than adsorption water and without perceptible emission of heat.

- **adiabatic**
a condition in which heat is neither gained nor dissipated. In this condition the vapour in an air-water vapour mixture becomes saturated without the addition or loss of heat.

- **adiabatic evaporation**
a constant pressure process of evaporation in farm buildings in which the air is held at a constant pressure and without addition of heat.

- **adiabatic lapse rate**
a lapse rate under adiabatic conditions, which is calculated as about 0.5°C temperature drop per 100 m of dumb for air saturated with water vapour and 1°C per 100 m climb for dry air.

- **adit**
  a nearly horizontal under inlet. It is sometimes referred to as a 'header' or 'drainage heading'.

- **adits**
  tunnels driven from the bottom or some distance downshafts or wells sunk in fissured and porous rocks; they intercept the water from the water-bearing fissures and lead it to the sump or well.

- **adjustability**
  the structural property of an outlet indicating the ease with which it can be adjusted for changed discharge, head, width, etc., without dismantling a major part or whole of the structure.

- **adjustable blade**
  a blade such as found on fans, the pitch of which can be changed or adjusted.

- **adjustable drawbar**
  the device on a tractor to which the load is attached and which may be moved vertically or horizontally to provide the most satisfactory loading arrangement.

- **adjustable gate**
  a device which is used for varying the size of an opening, as in a seed drill, in an irrigation ditch, or in 'gated' irrigation pipe.

- **adjustable hitch**
  the device capable of being moved vertically or horizontally to which a source of pulling power may be attached.

- **adjustable proportional module (apm)**
  a semi-module orifice outlet with adjustable roof block designed to distribute small variations in the parent channel, proportionately when set at the correct level relative to the depth of the parent channel.

- **adjusting nut**
  a burr or threaded device used to locate a given item at a desired setting.

- **adjutage**
  a tube inserted into an orifice; an effluxe tube.

- **adjuvants**
  any substance added to herbicides, such as wetting or spreading agents, which modify their physical characteristics and increase their killing power.

- **admiralty metal**
  a zinc bronze containing about 88 per cent copper, 10

per cent tin and 2 per cent zinc, much used for valves, fittings, gears, etc. It is strong and corrosion resistant.

- **admixture**
substances other than cement, aggregate, or water which are sometimes used as an additional ingredient for concrete, their purpose being to modify or improve it's properties in the desired respects so as to make it more suitable for the work in hand.

- **adsorbed water**
water in a soil mass, held by physico-chemical forces, having physical properties substantially different from absorbed water or chemically combined water at the same temperature and pressure.

- **adsorption**
a phenomenon consisting in the adhension, in an extremely thin layer, of the molecules of gases, or dissolved substances, or of liquids, to the surfaces of solid bodies, with which they are in contact, distinguished from absorption.

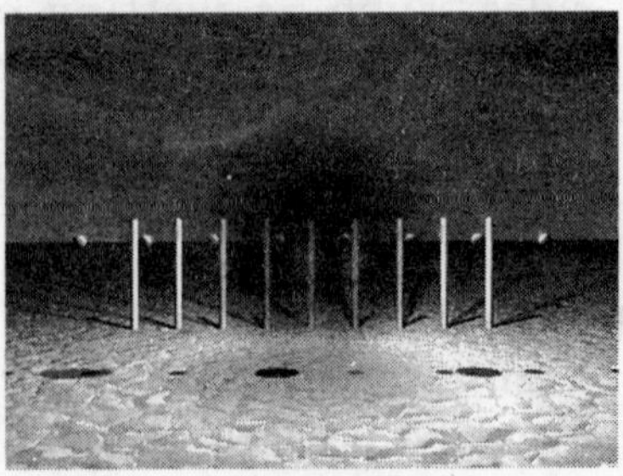

- **adsorption complex**
the group of substances in soil which are capable of adsorbing materials, the organic matter and colloidal clay forming the greater part of the complex and materials in silt and sand size exhibiting adsorption, but to a reduced extent in most soils.

- **adsorption water**
water held on the surface of solid particles by molecular forces with emission of heat.

- **advance curve**
advance curve is a relationship between 'distance travelled by a flow of water and unit of time' in surface irrigation methods.

- **advance phase**
is the portion of the total irrigation time during which water advances in over land flow surface from the upper end of the field to the lower end of the field. The time elapsed during this phase is called Advance time.

- **advance time**
it is time taken by water to travel the length, of an irrigation furrow.

■ **advance waste treatment**

waste water treatment beyond the secondary or biologic stage; it includes removal of such nutrients as phosphorus and nitrogen and a high percentage of suspended solids. Advanced waste treatment, known as tertiary treatment, is the 'polishing stage' of waste water treatment and produces a high quality effluent.

■ **advection**

the process of transport of an atmospheric property (such as heat, water vapour, or momentum) solely by the horizontal motions of the atmosphere.

■ **advective energy**

advective energy is the energy developed from horizontal heterogeneity in climatic parameters; or in simple words, it is the energy brought by winds. It is an important source of energy especially in tropical hot summers, which results in greater evapotranspiration than normally is due to the solar energy.

■ **adz hoe**

in forestry, a fire-trenching tool which resembles a carpenter's ADZ but has a broader blade and lighter head. Its tapered, rectangular eye is designed for a standard ship-carpenter's (double-curved) ADZ handle.

■ **aeolian deposit**

soil deposited in place by the action of the wind.

■ **aeolian soil**

soil formed from wind transported soil material.

■ **aerated grit chamber**

several makes of grit channel are aerated with coarse air bubbles so as to improve the separation of the grit from the organic solids.

■ **aeration**

the moving of (outside atmospheric) air through stored grain at low air flow rates to improve market quality for purposes other than drying. It is the process of mixing of air with soil or water in order to improve the supply of oxygen to plants and other organisms.

- **aeration drainage**
subsurface drainage primary purpose of which is to improve soil aeration, in order to satisfy the microbiological processes taking place in soil, and to change its physical and chemical properties in the direction of increasing soil fertility, to improve structure, increase the moisture capacity and water permeability of heavy clayey soil, accelerate the decomposition of the organic substances of peat soils, etc. At the same time, during periods of excessive moisture, at high ground water level, aeration drainage can produce a drainage effect.

- **aeration drying**
the process of forcing air through grain at a minimum rate of 3 cfm per bushel to remove excess moisture up to 15% above the desired level.

- **aeration equipment**
mechanical devices used in the cooling of stored grain.

- **aeration factor**
it is a ratio of oxygen consumption by soil microbes under optimum respiratory condition to the consumption in the presence of soil expressed in percentage of the optimal value. The limitation of this index is due to its artificial condition, which does not represent true natural condition of soil.

- **aeration irrigation**
a method of applying water by sprinklers, based on very low application rates; approximately one-half of the normal intake rate of the soil, combined with good nozzle pressure.

- **aeration porosity**
the pore space filled with air when the soil is under a tension of a 50 cm column of water.

- **aeration zone**
that portion of the lithosphere in which the functional interstices of permeable rock or earth are not (except temporarily) filled with water under hydrostatic pressure; that is, the interstices either are not filled with water or are filled or partly filled with water that is held by capillarity. The zone of aeration comprises the area where fringe water, intermediate water, and ground water may be found.

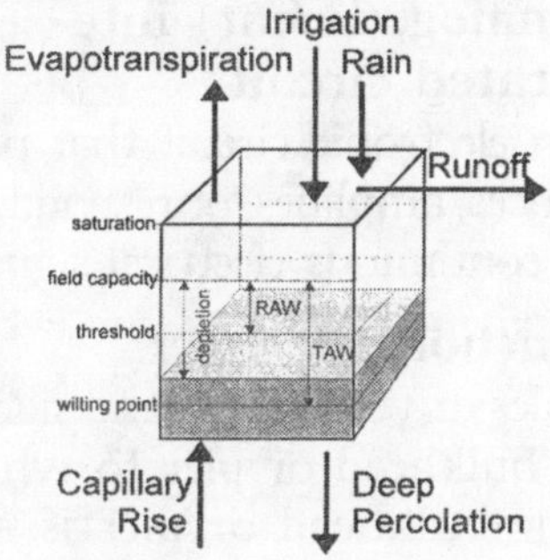

- **aeration, soil**
the process by which air in the soil is replaced by air from the atmosphere.
- **aerator**
a machine or other device to dissolve oxygen in sewage or water. It may also purge undesirable gases (hydrogen sulphide or carbon dioxide).
- **aerenchyma**
lysigenous gas spaces through which oxygen diffuses to supply submerged roots with oxygen.
- **aerial application**
the spreading of agricultural materials by air-craft.

- **aerial dusting**
dusting of pesticide, fungicide and herbicide powders from an aeroplane.
- **aerial irrigation**
the conveying of water under pressure through pipes to the area to be irrigated and passed out through nozzles of sprinklers.
- **aerial spraying**
application of pesticide or fertiliser in the form of spray by using aeroplane or helicopter, with the objective of covering vast area in a short time.
- **aerosol**
a can with contents under pressure.
- **aggregates**
soil units that contain mostly silt, clay, and sand particles which are held together by a gel type substance formed by organic matter.
- **agrarian**
which is based on agriculture.
- **agricultural education**
the application of engineering principles in agricultural settings.
- **agricultural mechanisation**
using machinery and power for producing fibre and food.

- **air intake system**
  the mechanism system that causes an air-fuel mixture or air to enter the combustion chamber during the intake stroke.

- **air-entrained cement**
  a kind of cement that contains additives which produce microscopic air bubbles in the concrete, thereby improving its resistance to damage by frost and freezing.

- **alloy**
  a substance that is a mixture of two metals having a unique property which is different from the original ones.

- **alternating current (AC)**
  an electrical current that reverses direction of flow periodically.

- **ambient temperature**
  temperature of the air around an object.

- **ammeter**
  a meter used for measuring current in an electrical circuit.

- **ampacity**
  the capacity of a conductor or other electrical device to carry current.

- **ampere (A)**
  the unit of measure of current in an electrical circuit.

- **analog (linear) integrated circuit**
  an electronic circuit that produces, amplifies, or responds to a continuous electrical signal.

- **anchor pile(s)**
  piles driven on the land side of a bulkhead or pier to which the bulkhead or pier is anchored or tied by rods, cables, chains, or other devices.

- **anchor plate**
  a piece of heavy sheet metal or wood embedded in or fastened to something solid and used to anchor or support two adjoining members.

- **anchored tree protection**
  a type of construction, in which branchy trees are placed against an eroding river bank, usually with their butts upstream, and are held in position by wire cables to anchors (posts, dead men etc.), in the river bank or weighted down by concrete blocks and wire cables or by stone mesh sausages.

- **ancillary shoring**
  the auxiliary shoring employed in the process of underpinning a wall; its purpose being to relieve the wall to be underpinned of as much load as possible.
- **ando soils**
  a zonal group of dark coloured soils high in organic matter developed in volcanic ash deposits.
- **Andrus, Leonard**
  early partner of blacksmith and inventor John Deere.
- **anemo**
  a Greek prefix meaning 'wind'.
- **anemometer**
  an instrument to measure the velocity of wind.
- **anemoscope**
  any device which is used to indicate and/or record wind direction.
- **aneroid barometer**
  an instrument used for measuring atmospheric pressure. It's main advantage over a mercury barometer is its portability.
- **angle bed digger**
  a mechanical device for digging potatoes on which the elevation mechanism for separating the potatoes from the loose soil is in two planes at an angle to each other.
- **angle blade scraper**
  an earth-moving device in which a blade is mounted at an acute angle from the direction of travel. The blades are usually adjustable in three planes, vertical pitch, horizontal angle and side tilt.
- **angle gate valve**
  a gate valve with an elbow cast on one end integrated with the body.
- **angle of current**
  in stream gauging, the angular difference between 90 degrees and the angle made by the current with a measuring section.
- **angle of external friction**
  angle between the abscissa and the tangent of curve representing the relationship of shearing resistance to normal stress acting between soil and surface of another material.
- **angle of hitch**
  the angle between the line of pull and the line of depth.
- **angle of internal friction**
  angle between the abscissa and tangent of the curve repre-

senting the relationship of shearing resistance to normal stress acting within a soil.

- **angle of lead**
the angle between the two vectors representing the waves of alternating voltage and current at any given instant. When the voltage passes through its zero value in a given direction before the current does, it is said to lead the current.

- **angle of offtake**
the angle formed between the centre line of flow through the offtake structure and centre line of the direction of flow in the parent channel; the angle being always measured on the downstream side of the junction of the lines.

- **angle of pull**
the included angle between the line of travel and a flexible connecting link (such as a chain) attached between the source of power and the load; for example, between the tractor and the pull-type plough.

- **angle of repose**
angle between the horizontal and the maximum slope that a soil assumes through natural process. For dry granular soils the effect of the height of slope is negligible; for cohesive soils the effect of height of slope is so great that the angle of repose is meaningless.

- **angle of sag**
the average angle that a sounding line makes with the vertical.

- **angle of twist**
see angle of offtake.

- **angle of wall friction**
see angle of external reflection.

- **angle valve**
a valve whose outlet is at right angles to its inlet branch thus combining in itself a valve and an elbow or bend.

- **angledozer**
a kind of bulldozer the blade of which can be inclined 60° to the direction in which the machine is moving.

- **angular perspective**
perspective in which some of the principal lines are oblique and some are not.

- **angular transformation**
before the analysis of variance technique is applied, the data having binomial distribution is transformed to normal distribution by making use of angular transformation; the data relating to percentages are transformed to degrees.

angularity
the angle between the thread of a stream or conduit and a line normal to the centreline of a structure (bridge, roadway, dam, etc.) crossing that stream or conduit.

angularity correction
the correction to be made to an observed velocity -when the direction of the current is not exactly at right angles to the discharge section line.

anicut
a dam or mole in a stream for maintaining and regulating irrigation.

aniline black
a black dye.

animal oils
lubricating oils for machinery obtained from animal tissues. Principal of these oils being sperm oil from whales and seals. Animals give an oil that does not dry and gum.

animal unit
livestock number equivalent to one mature cow (approximately 1000 lbs or 455 kg live weight), roughly one cow, one horse, one mule, five sheep, five swine, six goats.

animal unit month
measure of forage or feed required to maintain one animal unit for 30 days.

anion
a negative ion, one that moves to the anode in electrolysis.

anion exchange
replacement of acid radical ions such as NO_3^- or SO_4^{2-} in water by using io-=n exchange.

anion exchange capacity
the sum total of exchangeable anions that a soil can adsorb, expressed as milliequivalents per 100 grams of soil (or of other adsorbing material such as clay).

anion exclusion
clays being negatively charged attracts the positively charged cations and repels the negatively charged arsons as a consequence of electrostatic forces. The resultant reduction in concentration of anions near mineral surface is called anion exclusion or negative adsorption.

anisotropic mass
a mass having different properties in different directions at any given point.

- **annealing**
heating to, and holding at, a suitable temperature, followed by cooling at a suitable rate, for such purposes as (1) inducing softness, (2) improving machinability, (3) improving cold working properties, (4) obtaining a desired structure, (5) removing stresses and (6) facilitating diffusion processes. It is a heat treating process in which internal metal stresses are reduced, making the metal less hard and brittle.

- **anne's method**
a method for rapid determination of the quantity of organic carbon in soils. The technique uses attack by a boiling mixture of sulphuric acid and an 8 per cent solution of potassium bichromate, the chromic acid remaining at the end of the operation is titrated, making it possible to calculate the quantity of oxidised carbon. The results obtained differ little (7%) from those furnished by the dry method, and it is possible to make analyse in series. This method can be used not only to determine the organic matter content of soils taken from different places, but also to follow the development of a soil over a period of time. It can also be applied to samples of plant material.

- **annual closure**
the closure of an irrigation system for carrying out annual repairs.

- **annual depletion rate**
the average rate over a period of years at which withdrawals deplete the storage in a ground-water reservoir.

- **annual depreciation**
the amount of depreciation which occurs over a period of one year. Such depreciation is usually taken as a uniform quantity over the life of the structure, machine, etc., although it may be varied in accordance with a fixed rule.

- **annual fixed charges**
recurring annual charges in a sprinkler irrigation system including interest on investment, depreciation of equipment, and charges for insurance and taxes.

- **annual flood peak series**
the tabulation on graph in order of decreasing magnitude of the maximum peak floods of each year.

■ **annual inequality**
seasonal variation in the water level or tidal current velocity, more or less periodic, due chiefly to meteorological causes.

■ **annual load factor**
the load factor taken over an entire year.

■ **annual plant**
a plant that completes its life in one year or less.

■ **annual precipitation**
the total amount of water, usually expressed in inches of depth which is deposited on the earth in one year by rain, snow, hail, sleet, or other forms of precipitation; condensation of dew and frost is generally not considered as precipitation. More commonly called annual rainfall.

■ **annual storage**
1. the difference of storage in a reservoir at the beginning and end of a reservoir operations.
2.the difference between the maximum and minimum volumes in storage in a reservoir over a year of reservoir operation.

■ **annual use method of reservoir operation**
operation of a reservoir or a system of reservoirs aiming at using practically all the available storage each year without special regard for maintaining a minimum regulated flow.

■ **annual variation**
the general pattern of a particular element throughout the year, obtained by plotting the normal values of the element for each month and connecting the points by a smooth curve.

■ **annual weeds**
weeds having a life of one year only; they reproduce by seed only.

■ **annular drainage pattern**
drainage pattern in which ring like tributaries flow into the radial streams.

- **anomaly**

  a deviation from a norm for which an explanation is not apparent on the basis of available data.

- **anomorphic zone**

  that part of the lithosphere corresponding to the zone of flowage, within which simple compounds are built over into more complex ones, or reconstructed. The minerals thus formed are numerous. They have a high specific gravity and a complex structure.

- **anoxic**

  description of conditions without free oxygen. Oxygen then is available for respiration only from dissolved inorganic substances such as nitrate ions, $NO_3\sim$. It is not anaerobiosis. Anoxic sections in an activated sludge plant may be used for denitrification.

- **anstead's process**

  a process of composting wherein the materials to be composted are made into a heap of 6' x 3' x 3'. The material is mixed well and moistened with an 'activator' which may be of dung, urine, night-soil emulsion, ammonium sulphate or Adco. Rapid fermentation occurs and after about a week, the fermented, material itself constitute the activator or 'starter'. A quarter of the heap is then removed to form the basis of a second heap, and the first one is made to its original size with fresh waste, which is well mixed again, more activator being added. Fermentation takes place with rise in temperature and the process is continued. The advantage of this process is that once it has established itself, the external sources of activator are dispensed with, and the first heap itself constitutes the activator.

- **antecedent moisture**

  the degree of wetness of the soil at the beginning of a runoff period frequently expressed as an index determined by summation of weighted daily rainfalls for a 10 to 20 day period preceding the runoff in question.

- **antecedent precipitation index**

  a weighted summation daily precipitation amounts, used as an index of soil moisture. The weight given to each day's precipitation is usually assumed to be an exponential or reciprocal function of time with the most recent precipitation receiving the greatest weight. Such indices are used in estimating runoff and as measure of drought severity.

- **antecedent soil moisture**
  moisture content of soil prior to irrigation or at beginning of a run-off period.

- **antecedent stream**
  a stream which developed its course, are maintained it during and after a period when a disturbance such as a crustal movement, took place which altered or changed the drainage area.

- **antecedent temperature (ati)**
  the temperature occurring during some period prior to the defined event, where *i* is the period in months, days or hours, during which the temperature is totalled or averaged.

- **anthraciny**
  the breakdown of organic material by fungi and the further transformation of the results by passage through the alimentary canals of insects and worms, giving a dark coloured soil.

- **anthropic epipedon**
  mollic like horizon that has a very high phosphate corltent resulting from longtime cultivation and fertilisation.

- **anthropogenic, anthropic soils**
  changes in soils caused by people such as ploughing, fertilising, and using for construction.

- **anti corrosives**
  materials used in engines which resist corrosion. The materials usually contain active sulphur, phosphorus, or nitrogen.

- **anti foaming agent**
  a chemical added to an aeration tank to stop the production of foam. It usually consists of a mineral oil and a spreading agent.

- **anticaking agent**
  substance incorporated in or applied as surface treatment to granular fertiliser to prevent caking.

- **anticinal spring**
  a contact spring occurring along the surface out crop of an anticline, from a pervious stratum overlying one less pervious.

- **anticyclone**
area of relatively high pressure surrounded by closed isobars, the pressure gradient being directed from the centre so that the wind blows spirally outward in a clockwise direction in the northern hemisphere, counter clockwise in the southern hemisphere.

- **antidune**
sinusoidal-shaped bed features in phase with surface waves. These ridges usually move upstream.

- **antidune movement**
sediment waves on the bed of a stream which travel against the current. The downstream face of the sediment wave is eroded and the upstream slope receives deposit. They travel much faster than dunes. The water surface closely follows the profile of the antidune and the water surface waves appear to move against the current.

- **anti-flies**
a fence like device which is placed on the eaves of coops, hen houses, etc., to prevent domestic birds from flying to the roof.

- **anti-flocculent**
a material which when added to a liquid solution or suspension, will prevent sediment or other secondary material from separating out into small clumps or flakes.

- **antifriction bearing**
type of bearing in which moving parts are in rolling contact; ball, roller, or tapered roller bearing.

- **antipercolator**
the device in the carburettor that opens a vent when the throttle is closed, to permit release of fuel vapours in the high-speed circuit so that fuel vapour will not push fuel out of the high-speed nozzle.

- **anti-seep collars (soil and water)**
a collar or wall often placed around a drainage tube or line to prevent water seepage along the outside of the tube.

- **antitranspirants**
substances that retard or inhibit transpiration. These are of three types (1) apolar substances e.g., cetyl alcohol and oxyethylene decosanol, (2) plastics e.g., polyethylenes, and (3) substances that lead to stomatal closure e.g., phenyl mercuric acetate. A new group under the third category are alkenyl succinic acids, e.g., decenyl succinic acid.

- **ap**
surface layer of soil disturbed by cultivation or pasturing apron (soil engineering). A floor or lining constructed to protect soil surface from erosion below chutes, spillways or dams.

- **aperiodic**
a phenomenon not occurring regularly, but taking place at unequal intervals of time.

- **aperiodic motion**
motion that takes place in one direction only.

- **apparent cohesion**
cohesion of moist soils due to surface tension in capillary interstices. It disappears on immersion.

- **apparent density**
in soil mechanics the mass per unit volume of oven-dried soil, pore space included.

- **apparent ground water velocity**
the apparent distance covered by ground water in the saturated zone. It is defined as the volume of water per unit time divided by the cross-sectional area taken perpendicular to the streamlines.

- **apparent power**
the power in an AC electrical circuit calculated as the product of voltage times amperage.

- **apparent specific gravity**
the ratio of the weight of a unit volume of material to the weight of an equal volume of water under standard conditions. When dealing with soils the term may be applied to either oven-dry undisturbed field samples or to samples reworked in the laboratory.

- **apparent velocity (of ground water)**
the apparent distance moved in unit time by ground water. It is determined by $V=Q/F$ where F is the cross-sectional area perpendicular to the streamlines and Q the volume of water passing through this cross-section in unit time.

- **apple grater**
a device consisting of revolving knives fastened on a drum which grates apples, berries, grapes into very small particles preparatory to pressing the fruit juice.

application
general term for putting fertiliser on fields, crops, grassland, forests etc.

application factor
a factor applied to the concentration of toxicant in, e.g., a median tolerance limit (TLm) of a bio-assay, that is considered to make it safe. The factor can vary from a tenth to a hundredth of the 96-h or 48-h TLm figure. It is sometimes found by dividing the 96-h TLm by the maximum acceptable toxicant concentration.

application level
the height of aircraft above the ground or crop measured from the lower-most part of the airplane.

application rate
weight of fertiliser (for liquids often volume) applied unit area.

appropriate technology
it is a technology that may be suitable or proper in the context of a particular community, region, or country. It must be economically viable, technologically feasible and should fit in the socio-economic fabric of the local communities.

appurtenances
machinery, appliances, or auxiliary structures attached to a main structure, but not considered an integral part thereof, for the purpose of enabling it to function.

apron
1. a steel structure extending along the length of the cylinder in a roller gate at its bottom or top to effect seal or raise water level when the gate is at the lowest point of its travel.
2. the water bearing surfaces of roller and radial gates.
3. a floor or lining of concrete, timber, etc., to protect between a surface from erosion, such as the pavement below chutes, spillways, or the toe of dam.

aqua ammonia
solutions of ammonia absorbed in water and used primarily in ammoniating superphosphate and in making mixed fertiliser.

aqua regia
a mixture of one part of concentrated nitric acid and three

parts of concentrated hydrochloric acid. It is an extremely oxidising solvent capable of even dissolving metals like platinum and gold.

- **aquagel**

a gel-forming colloidal clay containing almost 100 per cent colloidal material, and consequently free from abrasive particles.

- **aquapedoturbation**

soil mixing by upwelling currents of water within the solum.

- **aquatic agriculture**

the management of natural and/or artificial means to increase the production of fish, oysters, or other animals or plants in fresh or salt waters.

- **aquatic weeds**

weeds growing in water, i.e., inside irrigation and drainage channels, ponds, lakes, rivers, storage reservoirs and all water containing conduits.

- **aqueduct**

1. a cross-drainage work in which the canal is carried over the drainage channel or natural stream and the bottom of the canal through, or the covering over the drainage openings, is above the high flood level in the drainage channel or natural stream.

2.a structure for conveying a canal or conduit across depressions or hollows. An aqueduct may be supported on piers and can be constructed in masonry, concrete, timber or steel in a trough or flume shape. The aqueduct may be connected as the ends to an earth or concrete channel.

- **aqueous**

a solution or suspension using water as the solvent. The term is often used to describe a gaseous compound dissolved in water, e.g., aqua ammonia, a form of ammonia obtained by dissolving in water. A number of fertiliser solutions in which water is used as a solvent are also termed aqueous fertiliser solutions.

- **aquic moisture regime**

these soils are water saturated for at least enough time (several days) so that reducing conditions exist. Low chroma mottles normally are indicative of this condition.

▪ **aquiclude**
a formation which although porous and capable of absorbing water, does not permit its movement at rates sufficient to furnish an appreciable supply for a well or spring.

▪ **aquifer**
a geologic formation that transmits water in sufficient quantity to supply the needs for a water development. Aquifers are usually saturated sands and gravels and fractured, porous, cavernous, and vesicular rock.

▪ **aquifuge**
a formation which has no interconnected openings and hence cannot absorb or transmit water.

▪ **arable**
capable being ploughed. Good arable land is that which has already been cultivated.

▪ **arable crops**
these are crops which require cultivation.

▪ **arable farming**
farming system that involves the production of crops requiring tillage.

▪ **arable land**
land which is suitable for ploughing. The term also applies to land, which is actually ploughed and cultivated under the crops.

▪ **arbor press**
an appliance used for forcing arbores or mandrels into or out of work by the aid of a screw press of hydraulic means.

▪ **arboriculture**
intensive cultivation of individual trees, possibly for fruits, mats, gums, resins etc.

▪ **arch**
a curved structure supporting a weight of material above, it. For example, an arch dam or the curved top of conduit.

▪ **arch abutment**
the abutment of an arch element is the surface, at either end of arch, which contacts the rock of the canyon wall. Arch loads are transferred through the arch abutments to the canyon walls.

▪ **arch bridge**
a bridge in which the main supporting elements are arches.

▪ **arch centre**
a temporary structure for supporting an arch while in the process of construction.

- **arch centre line**
  an arch centre line is the locus of all median points of the thickness of an arch section.

- **arch dam**
  an arch dam is a solid concrete or masonry dam, curved upstream in plan, which, in addition to resisting part of the pressure of the reservoir by its own weight, obtains a large measure of stability by transmitting the remainder of the water pressure or load by arch action into the canyon walls or other abutments.

- **arch dam in series**
  a series of arch dams placed one behind the other in descending height, the space between successive arches is filled with water to the level of the crest of the lower arch; by this means, the pressure on each arch would be limited to that due to the difference in height of two successive arches; consequently, the arches would be of constant thickness.

- **arch element**
  an arch element, or arch, is a portion of a dam bounded by two horizontal planes at distance of one unit apart.

- **arch gravity dam**
  a curved solid masonry dam which depends upon both arch action and gravity action for stability.

- **arch section**
  a section representative of an arch dam as it would appear if cut by a horizontal plane.

- **arched culvert**
  a culvert having arched roof supported on piers or abutments.

- **archimedean drill**
  a hand drill for light work, whose stock is formed into a multiple spiral or quick pitch over which a nut or spool slides easily. The act of sliding the nut, up and down stock makes the latter and consequently the drill, rotate first one way and then the other, pressure being applied by a knob or ball at the opposite end of the stock.

- **archimedean screw**
  a device, attributed to Archimedes, consisting of a tube bent spirally around an axis or alternatively broad threaded screw encased by a hollow open-cylinder, and used to raise water by rotating the apparatus when partly immersed in a slantwise direction.

- **arching**
the transfer of stress from a yielding part of a soil mass to adjoining less yielding or restrained parts of the mass.

- **area**
the amount of surface included between certain closed boundary lines; any particular extent of surface region or tract.

- **area curve**
a curve which expresses the relation between area and some other variable, such as cross-sectional area of a stream against water surface elevation, surface area of a reservoir against water surface elevation, etc.

- **area flowing full**
the state of flow when the fluid occupies the whole area designed to accommodate it.

- **area grouting**
grouting to be done in an area to stop water passage through the disintegrated rock and also to increase the bearing strength of the strata. It is comparatively shallow and low pressure grouting.

- **area irrigated**
the area to which water has been applied.

- **area leased**
area for which cultivators have executed an agreement for taking canal water for certain periods.

- **area mature**
the area irrigated upon which crops have matured.

- **area non perennial**
the area served by a non perennial canal.

- **area of artesian flow**
a land or water surface which lies below a piezometric surface. It is an area in which the water of some underlying aquifer is under sufficient pressure to rise above the surface.

- **area of gas lift flow**
an area in which the ground water, if tapped, rises to the surface because of the presence of gas.

- **area of influence**
the area is plan covered by the draw down curves of a given pumping well or combination of wells at a particular point. The area of influence for a given rate of discharge may vary with the period of withdrawal and the rate of recharge.

- **area perennial**
the area served by a perennial canal.

- **area sampling**
  area sampling is an application of cluster sampling with other design features interwoven, the idea being that by drawing probability samples of specific areas and sampling appropriately within them, it is possible to obtain a probability sample of the population.
- **area time equivalency ratio**
  it is the ratio of number of hectare-days required in monoculture to the number of hectare-days used in intercropping to produce identical quantities of each of the component crop.
- **arenaceous**
  from the Latin word 'arena' meaning sand. Like, or pertaining to sand. An example is arenaceous limestone, or a sandy limestone.
- **arenic horizon**
  this is a loamy fine sand or coarser horizon more than 50 cm (20 in.) thick over an argillic horizon.
- **arenicolous**
  growing in sand or sandy areas; living in or on sand.
- **areometer**
  a hydrometer.
- **argillaceous**
  containing or consisting of clay. An example is argillaceous limestone, meaning one containing a high percentage of clay.
- **argillic horizon**
  an alluvial horizon of silicate clay accumulation.
- **argillipedoturbation**
  It is mixing of materials in solum by mass washing movements of expansible clays.
- **argillisation**
  the transformation of a rock or a mineral into clay.
- **arid**
  a term applied to regions of climates that lack sufficient moisture for crop production without irrigation.
- **arid climate**
  a generally extremely dry climate with an annual average precipitation usually less than 250 mm and generally rainfall is well short of evapotranspiration demand of the atmosphere.
- **arid erosion**
  a form of erosion or general wearing away of rocks which takes place in arid countries, such erosion being due largely

to the wind. Arid erosion is equivalent to 'desert erosion' and is in contradiction to normal, glacial and marine erosion.

- **arid lands**
area where precipitation is insufficient for crop production.

- **arid region**
area where the potential water losses by evaporation and transpiration are greater than the amount of water supplied by precipitation.

- **arid zone**
the area where total annual rainfall is usually less than 10 inches. The desert and semi-desert regions where there is only a sparse vegetation, come under this zone. This area is not suitable for crop production unless irrigation facilities are provided.

- **aridic or torric moisture regime**
these soils are both dry more than half the time when not frozen and never moist more than 90 consecutive days when soil temperatures are above 8°C at 50 cm depth.

- **aridity**
1. area having less than 250 mm of rains per annum,
2.area where the total evaporation potential exceeds the actual precipitation.

- **armature**
the rotating component of a DC electric motor.

- **aromatic compounds**
organic compounds formed from benzene rings. They may be not easily biodegradable and with attached chlorine atoms they often form the base of pesticides, antiseptics, etc.

- **aromatic solvents**
coal tar and petroleum derivatives.

- **arrow cutter**
an implement shaped like an arrow and consisting of a cutter, a tractor and rudder.

- **arrow head points**
a type of cultivator shovel usually narrow with a flared point.

- **arrows**
V-type block kept with their vertex on upstream side in the path of flow.

- **arroyo**
a stream channel or gully usually rather small with steep banks, and dry much of the time.

- **arterial drainage**
  a system of drainage in which the flow from a number of branch drains is led into one main channel.

- **artesian aquifer**
  a formation of permeable material which will yield an appreciable quantity of water and in which the water is under pressure.

- **artesian basin**
  a subsurface water basin where the water is under pressure, because of overlying confining material. The water will rise into or above the confining material if given an opportunity to do so.

- **artesian capacity**
  the rate at which a well will yield water at the ground surface as the result of artesian pressure.

- **artesian discharge**
  the discharge of water from a flowing well. The rate of discharge of water from a flowing well.

- **artesian flow**
  1. flow from a well produced by artesian conditions.
  2.subterranean flow of water, often under considerable pressure head.
  3.the flow from an underground water-bearing stratum with sufficient pressure to force water above ground level.

- **artesian flow area**
  the land or water surface area within which the water of some underlying aquifer is under sufficient pressure to rise above the surface of the ground.

- **artesian flowing well**
  artesian well tapping confined, water under pressure sufficient to cause it to rise above the land surface at the well.

- **artesian head**
  pressure head of artesian water.

- **artesian leakage**
  the slow percolation of water from artesian formations into the confining materials of a less permeable but not strictly impermeable character. Such percolation causes a reduction in artesian pressure, depending on the relative impermeability of the materials in the confining formations.

- **artesian pressure**
  pressure of the artesian water.

- **artesian pressure surface**
in ground water hydrology, a piezometric surface above the upper surface of the zone of saturation.

- **artesian reservoir**
a formation or aquifer in which the water is under artesian pressure.

- **artesian spring**
a spring whose water issues under artesian pressure, generally through some fissure or other opening in the confining bed that overlies the aquifer.

- **artesian waste**
the waste of water from artesian wells, due to flow at the ground surface when the water is not used and the wells are allowed to flow without being capped or shut off, the wells being abandoned or temporarily out of use, or below the ground when the wells are not cased or when the casing is in a leaky condition.

- **artesian water**
subsurface water under sufficient pressure to cause it to rise above the bottom of some fissure or other opening in the confining bed that overlies the aquifer.

- **artesian waterpower**
the power developed at the mouth of an artesian flowing well by the pressure of the water discharged from the well.

- **artesian well capacity**
the rate at which a well will yield water at the surface as the result of artesian pressure.

- **artesian well meter**
a type of meter to measure water in flowing artesian wells.

- **artificial chemical control measures**
chemical control measures by increasing nitrogen in the water by immersing bark scrap, tapioca or other decomposable matter, addition of cut grass, or changing salt content of water.

- **artificial classification**
a classification based at least in part on artificial characters.

- **artificial drainage**
drainage instituted by man.

- **artificial navigable waterway**
a stream, channel or canal made navigable by dredging and/or construction of controlling works.

- **artificial nourishment**
  the process of replenishing a beach material by artificial means; e.g., by the deposition of dredged materials.

- **artificial rainfall**
  water artificially applied in the form of rain, usually by a rain maker, for experimental purpose.

- **artificial recharge**
  ground-water recharge thro-ugh works provided, practices followed, such as spreading, or artificial changes made in natural conditions, primarily for that purpose.

- **artificial subsurface irrigation**
  application of irrigation water below surface of soil with perforated or porous pipes laid underground below the root zone and water led into the pipes by suitable means. Method may lead to soil getting saline or alkaline and the neighbouring land getting damaged due to seepage.

- **artificial watercourse**
  a surface watercourse constructed by human agencies.

- **artiquar**
  tillage after burning of land.

- **arviculture**
  crop science.

- **ash**
  the non-volatile residue resulting from complete burning of organic matter.

- **ash (volcanic)**
  loose, fragmental material, the particles being largely less than 4 mm in diameter, blown from a volcano; includes volcanic dust.

- **ash analysis**
  a technique for the study of plant nutrition which consists in subjecting the plant material to a high temperature (about 600°C) and analysing the ash content. Only the inorganic elements can be detected. Carbon, hydrogen, oxygen and nitrogen cannot be detected by this method.

- **ashford shutter**
  a type of falling shutter in which a strut (on the downstream) is used to support the shutter, which can be released by a lever. There is a separate strut for each shutter which itself revolves round the hinge at the bottom. The whole shutter system falls automatically if one lever is released. Slightly projecting the lever beyond the end of the

shutter and providing a hinge in the middle of the lever accomplish this.

- **ashing**
the process of burning resulting in the breakdown of complex compounds into simpler ones.

- **ashy**
having the consistence and the colour of ash. The A_2 horizons, which have been greatly weathered by leaching waters in an acid environment is usually ashy.

- **aspect**
the compass direction towards which land slope faces. The direction is measured down slope and normal to the contours of elevation.

- **asphalt concrete**
a concrete made by adding proportionate amounts of asphalt to the concrete mixture.

- **asphalt facing**
a relatively thick, heavy-duty asphalt surfacing designed to resist severe erosion, abrasion, high water pressure, and in some instance, ice pressure. A facing may have the additional purpose of acting as an impervious layer to prevent leakage through the structure.

- **asphalt impregnated felt**
a roofing and waterproofing material consisting of saturated asbestos or rag felt, cemented together with asphalt or tar pitch.

- **asphalt joint sealer**
an asphalted mastic used to seal joints in masonry construction, i.e., the joint between a masonry well and a concrete floor.

- **asphalt mattress**
these are terms applied, according to size, to prefabricated impervious flexible units composed of a mastic type of mix, suitably reinforced if required.

- **asphalt membrane**
a layer formed by an application of a relatively hard, high viscosity, high-softening-point asphalt, usually without filler or aggregate. It is usually covered or buried to insure its protection and to prevent weathering.

- **asphalt revetment**
a protective surfacing that has the principal function of preventing erosion by water. The term 'subaqueous revetment' refers to that part of a revetment placed under the

surface of the water 'upper bank paving' is that part placed above the surface of the water.

- **asphalt shingles**
a manufactured roofing material, composed of an asphalt saturated fibre base covered with a layer of mineral granules.

- **asphyxia of the soil**
lack of respiration in soil for want of oxygen. It usually occurs under the effect of an influx of water in the soil, which increases the proportion of carbon dioxide at the expense of the oxygen. This adversely affects plants and animal micro-organisms of the soil.

- **aspidisca**
a free-swimming ciliate protozoon that is often found in trickling filters and sometimes in activated sludge, especially if the effluent is of high quality.

- **aspirated psychrometer**
psychrometer with artificial ventilation of two thermometers (dry bulb and wet bulb) like Asmann Psychrometer or whirling Psychrometer.

- **aspirating**
a method of cleaning grain by drawing air through the grain by connection to the intake of a fan.

- **aspiration**
it is the process of cleaning by air blast and separating the foreign material which is substantially lower in specific gravity than the produce to be cleaned.

- **aspirator**
a device resembling an ejector, in which water passes through a nozzle, whose outlines conform to the *vena contracta*. This device induces a suction current in a connection pipe and exhausts the air thus creating a high vacuum.

- **assembly drawing**
a drawing of the structure or machine put together, showing the relative positions of the different parts.

- **assessable land**
any land located within the boundaries of an irrigation, drainage, sewerage, or district of similar character, which is subject to assessment to defray the cost of constructing improvements in the district. Such lands usually include all those within the district boundaries except those specifically exempted by statute, the latter class generally in-

cluding district right of ways and other property, and property publicly owned for public purposes.

- **assessment**
the charge against any particular parcel of land within the boundaries of an irrigation, water, sewer, drainage, or other district created for the purpose of constructing improvements, or a share of the total cost of such improvement, usually based upon the proportionate benefits received by such parcel as a result of the improvement. A special assessment is direct tax levy assessed against property to pay for improvements, which ordinarily are a direct benefit to the property itself.

- **assessment of irrigation charges**
process of determining amounts to be paid for project services by groups of beneficiaries and individuals.

- **assignable causes**
the variations due to other than chance or random causes, where the basic conditions and, thus, the population have changed, reflect the presence of assignable causes.

- **Assimilation**
a general term for the whole of the metabolic processes by means of which food is built up and utilised by the plant. Often used as equivalent to photosynthesis.

- **assimilative capacity**
of a receiving water, its ability to receive discharge of wastes without suffering pollution.

- **assist feed mechanism**
the seed metering device on a semi-automatic potato planter which enables the operator to assure that it functions properly.

- **assmann psychrometer**
a hand instrument used for measuring the relative humidity of air at different heights in the observatory or in crops. Out of the two thermometers in this instrument, bulb of one is made wet, air is drawn and the instrument is held at different heights as measured by graduated pole driven in the observatory.

- **associated cost**
the value of goods and services needed, beyond project cost, to make the output of the project available.

- **associated cropping**
  synonymous with inter-cropping.

- **association**
  1. the degree of dependence or independence, which exists between two or more variants whether they be measured quantitatively or qualitatively.
  2. from ecological point of view, an assemblage of plants having similar environment requirements and including one or more dominant species, from which species the assemblage derives a definite aspect.

- **astragal**
  a small moulding of circular section. A moulding, attached to one of a pair of swinging doors, against which the other door strikes.

- **asymmetrical**
  without proper proportion of parts.

- **asymmetrical (nose) cut water**
  a cut-water having asymmetrical nose shape to suit special flow conditions, sometimes, warranted by cross-currents and oblique Sow.

- **asymmetrical distribution**
  a distribution which is not symmetrical, that is to say, for which there is no central value *a* such that f (x—a)=f(a—x), f(x) being the frequency function.

- **at the base**
  a term which means at the ground level while measuring the diameter of a standing tree as contrasted to the term at the stump or at breast height.

- **atmometer**
  an instrument for measuring evaporation.

- **atmospheric condensation**
  when moisture laden air is uplifted and cooled sufficiently, small droplets of water are formed as in a cloud or a fog. The condensation takes place on dust particles.

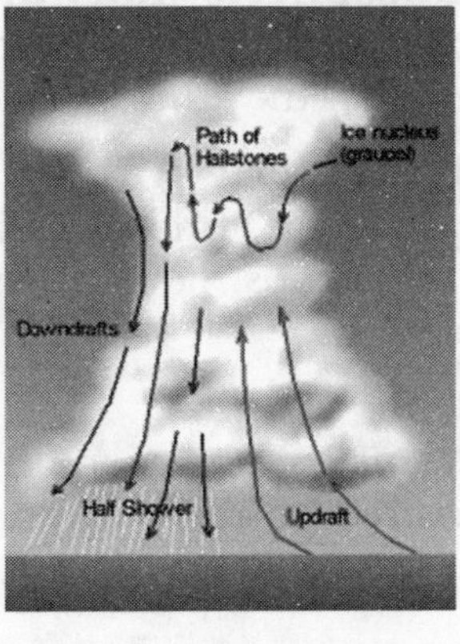

- **atmospheric conditions**
  usually refers to temperature, relative humidity, motion, cloudiness, barometric pressure and suspended matter such as dust, smoke, odours, etc.

- **atmospheric drum dryer**
  a dehydrator for liquid food products consisting of a large, steam-heated drum which is coated with the food product as it slowly revolves. The water content is rapidly evaporated and close-fitting knives scrape the dehydrated material of the drum.

- **atmospheric induction engines**
  any internal combustion engine using air at atmospheric pressure as compared to an engine equipped with a supercharging device on the air intake.

- **atmospheric moisture**
  water as it occurs in various forms in the atmosphere.

- **atmospheric water**
  water in the atmosphere in either gaseous, liquid, or solid state.

- **atom**
  the smallest unit of an element that retains the properties of the element.

- **atomic absorption spectrophotometer**
  an instrument widely used for measuring trace concentrations of elements in water. A light source with a wavelength readily absorbed is directed through a flame in which the sample is being atomised. The intensity of light absorbed is related to the concentration of that element in the sample. The method is fast and generally free from interference between elements, but it requires a different light source for each element.

- **atomisation**
  the process of generating a population of free atoms required for analytical estimation using a flame photometer and atomic absorption spectrophotometer. The process in which energy required for atomisation is supplied by the flame is called flame atomisation. There are five stages in flame atomisation: (1) Nebulisation (2) Droplet precipitation (3) Mixing (4) Desolution (5) Compound decomposition. Non-flame atomisation is more efficient in producing free atoms.

- **atomiser**
one of the important components of the atomic absorption spectrophotometer and flame photometer used in fertiliser analysis. The function of an atomiser is to generate the population of atoms from the analytical solution.

- **atritor flash dryer**
a device for flash drying of sludge, with a fast moving rotor carrying pegs that pass between other stationary pegs on a casing with hot gas flowing through it. The dried sludge is therefore pulverised. The dryer normally accepts only dewatered sludge.

- **attached ground water**
the portion of ground water adhering to the pore walls. It is assumed to be equal in amount to the residual water left after drainage and is measured by specific retention.

- **atterberg flask**
a graduated flask with a lateral siphon, used in grain size analysis to extract successively the suspension of increasingly fine particles.

- **attic ventilator**
usually openings in gables or ventilators in the roof. Also, mechanical devices to force ventilation under a roof by use of power driven fans.

- **attracting groin**
a groin which attracts the current towards itself and fixes the deep channel close to itself.

- **attrition**
1. the wearing down of the monolayer by continual slight impairment, as from winds, microbiology of the reservoir.
2.act of rubbing together, friction, wearing, abrasion.

- **attrition mill**
a mill which grinds by the method of friction or rubbing action to reduce the size of particles. Used in feed and flourmills.

- **a-type frame**
a rape of building framework resembling the capital letter Λ; usually built by using two common rafters which extend from the peak to the floor of the building.

- **auflage humus**
a kind of pest which is looser and usually less acid and more fertile than torf.

- **auger conveyor**
a materials handling device used for horizontal or vertical movement of free flowing

materials. A rotating screw, usually within a trough or cylindrical tube, moves the material.

- **auger fertiliser distributor**
  a device used for application of dry commercial fertilizers which utilises an adjustable speed auger as a metering mechanism.

- **auger hole method**
  in a test for soil permeability, an auger hole is bored into the ground below the water surface. The water is pumped from the hole, and as the hole refills, the rate at which the water rises is observed, and the permeability is found from this rate.

- **auger soil**
  a tool for boring holes into soil. A soil auger is used to withdraw a small soil sample for observation.

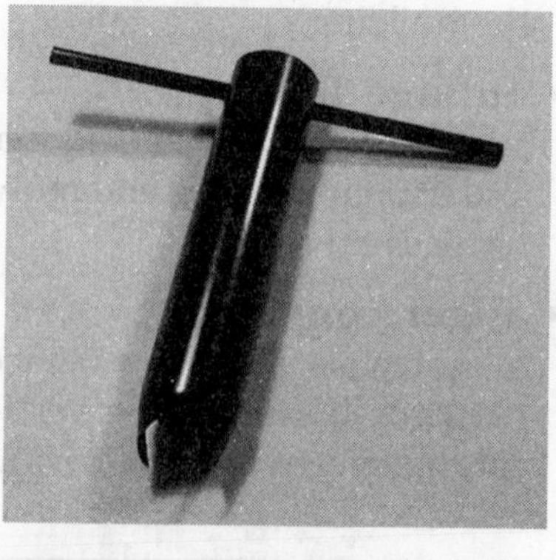

- **augmentation cropping**
  it means growing of a crop to supplement the yield/monetary returns accruing from the main crop. For example growing of forage brassica to augment the fodder yield of berseem in first cut.

- **austausch**
  measure of degree of turbulence within a turbulent zone, in terms of exchange of mass across unit horizontal surface during unit of time; so named from German word meaning interchange.

- **authicenic**
  formed in place, at the expense of the minerals or organic substances of the site itself.

- **autoclave**
  an airtight vessel into which superheated steam can be introduced under pressure, at temperatures above 212°F. It is used for sterilising.

- **autocology**
  the study of the details of how an individual or a species interacts with its environment.

- **autocorrelation**
  the internal correlation between members of series of observations ordered in time or space.

- **autogenetic drainage**
drainage due to erosion caused by the waters of the constituent streams.

- **autogenetic topography**
conformation of land due to the physical action of rain and streams.

- **autogenic factors**
dominating factors of change which are due only to the individuals in a plant community, e.g., shade, root, competition, etc.

- **autolysis**
the disintegration of cells by the action of their own enzymes, part of the endogenous phase of growth.

- **automatic choke**
a choke that positions the choke valve automatically in accordance with engine temperature.

- **automatic constant flow offtake regulator**
an ungated structure yielding constant flow on the module principle.

- **automatic controls**
any set of switches, valves, etc., which are operated without manual assistance when certain preset conditions are reached.

- **automatic draft control**
a mechanism, usually hydraulically actuated, in which the control level position represents given implement draft. Variations in draft cause the implement to be automatically raised or lowered to restore the pre-selected draft.

- **automatic flap gate**
a type of check valve. A device located on the end of the discharge pump or main line, which prevents backflow.

- **automatic gate**
a gate that operates automatically through unbalanced water pressure, or by electricity.

- **automatic pickup hay baler**
a hay baler with a pickup cylinder that delivers the hay directly to a floating feed cross auger.

- **automatic planter**
a simple bullock drawn automatic seed-cum-fertiliser planter, useful for metering correct quantity of seed and for placing fertiliser at proper depth. Four such planters can be attached to tool bar of tractor. Also equipped with a pair of marker arms for aligning the rows.

- **automatic position control**
  a type of hydraulic control system in which there is a definite relation between the position of the implement and the position of the control lever, thus allowing the hydraulic cylinder to move an implement automatically to a pre-selected position or depth as determined by the control lever.

- **automatic potato planter**
  a device used in planting potatoes consisting of a picker wheel to which are attached from three to twelve picker arms. These arms pass through the picking chamber containing the seed. The seed is picked up and dropped down a seed spout, which guides it into the ground.

- **automatic sampler**
  a device which removes a representative portion of a falling stream of grain without attention from an operator. Also catches a portion of water flowing by, as in a stream.

- **automatic semi-constant flow offtake regulator**
  an ungated structure working on semi-module principles with constant flow at designed upstream water level.

- **automatic shutter**
  a shutter which maintains required headwater level with all discharges.

- **automatic spillway**
  a spillway which operates through the rise of water in the reservoir or canal, generally by lowering the spillway crest or increasing the area of the outlet orifices.

- **automatic spillway gate**
  a gate which responds to flood levels without manual control.

- **automatic sprinkler**
  a system of overhead piping connected to the water supply, with heads or nozzles which open at a prescribed temperature for fire protection of the interior of a building.

- **automatic station**
  a hydroelectric generating station wherein the generators are automatically started, synchronised, and connected to the electric distributions system, and again disconnected and shut down; the operation being governed by load or water conditions as desired.

- **automatic tide gauge**
an instrument consisting of a float connected to a recording drum which automatically records tidal heights with respect to time.

- **automatic valve**
a valve which opens or closes when prescribed conditions are reached, without human assistance.

- **automatic water stage recorder**
a permanent attachment installed over an observation pipe for continuous record of ground water level with a recording drum, run by a clockwork.

- **autophyte**
a plant which synthesises complex food substances from simple compounds, e.g., green plants.

- **autoradiography**
a technique whereby a thin slice of tissue containg a radioactive isotope is placed in contact with a photographic plate. The image obtained on development shows the distribution of the isotope in the tissue.

- **auxanogen**
a piece of apparatus used to measure the elongation of plant shoots. It incorporates a lever to exaggerate the actual growth.

- **auxanometer**
an instrument which is used to measure growth of plants. These magnify the growth in length of an organ so as it could more easily be read.

- **auxiliary engine rotary plough**
a rotary plough employing an internal combustion engine as a power unit to operate the mechanisms. The engine is mounted on the machine, which is not usually self-propelled.

- **auxiliary flood ditch**
a ditch used to remove surplus surface or runoff water.

- **auxiliary gasoline engine**
a gasoline-burning engine serving to supplement or take the place of another power source; as an engine on a machine to supplement a ground drive mechanism or another engine.

- **auxiliary projection**
a projection of an objective on a plane which is perpendicular to one of the principal planes of projection and inclined to the other two.

- **auxiliary reservoir**
  a reservoir which supplements the storage of a main reservoir.

- **auxiliary section**
  a sectional view, projected upon an auxiliary plane, to show the shape of a surface cut by a plane, or to show the cross-sectional shape of an object.

- **auxiliary siphon type water seal**
  a water seal made effective by the provision of a baby siphon placed either in front or by the side of the main siphon. The two siphons are connected through a pipe conduit.

- **auxiliary spillway**
  a dam spillway which serves to carry runoff in excess of that carried by the principal spillway.

- **availability factor**
  the ratio of the time a piece of equipment or a unit is ready for or in service to the total time interval under consideration.

- **available heat**
  in drying grain, the quantity of heat in air, that, with proper exposure, can be utilised in evaporating water from grain. It is the sensible heat before such exposure minus the sensible heat after exposure. It is expressed in Btu per lb. or Btu per cu. ft.; used in connection with velocity of airflow for conditioning of harvested produce.

- **available moisture capacity of the soil**
  the amount of soil moisture that may be available in the wetted zone following an irrigation is the difference between the field capacity and the permanent wilting point.

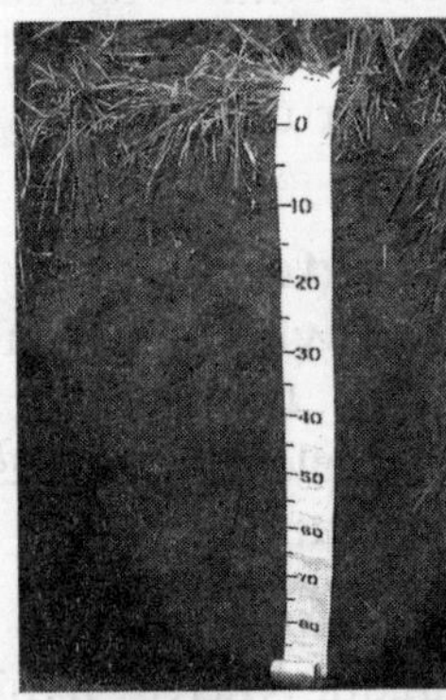

- **available nutrients in soils**
  a part of the plant nutrient in the soil that can be taken up by growing plants immediately. Available nitrogen is defined as the water-soluble nitrogen plus the part that can be made soluble or converted

into free ammonia. Available phosphoric acid is that part, which is soluble in water or in weak dilutes acid such as 2 per cent citric acid. Available potash is defined as that portion which is soluble in water or in a solution of ammonium oxalate.

- **available oxygen**
  the quantity of atmospheric oxygen dissolved in the water of a stream. It is the quantity of dissolved oxygen available for the oxidation of organic matter.

- **available pressure**
  the static pressure available for the operation of hydraulic equipment such as the pressure to operate an irrigation sprinkler.

- **available pressure head**
  in case of an inverted water discharge, openings of which are in a bed that is permeable but not saturated, it is the vertical distance between the level at-which the water to be drained off is admitted, and the average level of the discharge openings.

- **available soil moisture**
  the amount of water in the soil at any time in excess of the wilting coefficient, expressed either as percentage by weight of dry soil, or as equivalent of water per unit depth of soil.

- **available working head**
  the minimum difference available between supply and delivery water levels or headwater and tail-water.

- **avalanche voltage**
  the reverse-bias voltage required to cause a zener diode to conduct.

- **average annual flood**
  the mean of the annual floods over a number of years. The smallest of the annual floods during the period of record is called the 'minimum annual flood'.

- **average holding**
  figures arrived at after dividing the total area of holding by the total number of agricultural holdings.

- **average infiltration rate**
  the rate obtained by dividing the mass infiltration, that occurs during the time rainfall intensity exceeds the infiltration capacity, by the time during which infiltration occurs at capacity rate.

- **average life**
  an appraisal term defined as the mean or normally expected usable life of a property.

- **avogadro's law**
equal volumes of all gases contain the same number of molecules at the same temperature and pressure.

- **avogadro's number**
the number of molecules in a gram molecule of gas (or atoms in a gram-atom of the gas) at S.T.P. is 6.0225×10^{23}.

- **avulsive cutoff**
the term is used to denote the river's action when an avulsion takes place; also, briefly, to describe the area of land circumscribed by the old and the new channels.

- **axial piston pump**
a kind of hydraulic pump where the pistons are parallel to the pump's drive shaft.

- **b**
this symbol is added to the designation of a buried genetic horizon or horizons. Horizon of another solum may or may not have formed in the overlying material, which may be similar to, or different from, the assumed parent material of the buried soil.

- **b horizon**
soil below the horizon or topsoil, which is generally referred to as subsoil.

- **b/abc soil**
a soil three horizons which has undergone intense upward movement of solutions by capillarity.

- **ba**
Symbol for the biological activity of a soil, measured by the rate of humification of cellulose and albumin.

- **babbits**
applies to alloys high in tin, having general composition as copper 4.5 tin 91.0 and antimony 4.5, because of the pioneer work of Isaac Babbit. These are used as bearing metals.

- **back furrow**
the resulting ridge of soil turned up when a furrow slice is lapped over the first furrow slice when starting the ploughing operation.

- **back ploughing**
ploughing to sufficient depth to control regrowth.

- **back pressure**
the pressure on the opposite side of the piston, which opposes a resistance to the working stroke.

- **back siphonage**
water sucked out of vessels due to a low pressure in the supply pipe.

- **back to back setup**
a technique of accelerated testing of farm tractors. A test tractor is loaded by a tractor load machine through direct tire-to-tire contact. The external forces on the test tractor are similar to those found in the field, i.e., forces at the axles are forward and forces at the drawbar are backward.

- **back washing**
a method of well development based on the principle of surging or agitating the formation at the well by hydraulic or water pressure or air pressure to prevent bridging of the sand particles and to remove a large portion of the finer material.

- **back wiring**
a method in which a circuit conductor to an electrical device is connected by inserting the conductor into a spring-loaded hole in the device.

- **backbiowing**
reversal of the flow of water under pressure in a well to free the screen or strainer and the aquifer of clogging material.

- **backfire**
an explosion in the exhaust pipe or muffler of an internal combustion engine. Usually caused by incomplete combustion in the cylinder. In oxyacetylene welding equipment, the loud snapping or pop caused when the flame goes out suddenly. Usually caused by some interference in tip of the welding equipment.

- **backfire test**
one of seven tests specified by the Society of Automotive Engineers to evaluate air cleaners.

- **backflow**
in plumbing, the flow of water or other liquids into the distributing pipes of a potable water supply from any source or sources other than its intended source. The backing up of water through a conduit or channel in the direction opposite to normal flow.

- **backflow preventer**
a device for a water supply pipe in a building, to prevent the backflow of water into the water supply system from the connections on its outlet end.

- **backing**
a change of direction of the wind in an anti-clockwise direction.

- **backsight**
the level rod reading on a point of known elevation. A backsight is a plus sight and is always added to the elevation to obtain the height of instrument (H.I.).

- **backslope**
the less sloping side of a ridge. Contrasted with 'escarpment' the steeper slope.

- **bacteria bed**
a trickling filter.

- **baffle aerator**
an aerator wherein baffles are provided to cause turbulence of flow.

- **baffle fall**
a fall or drop having a baffle wall or baffle below the notch or orifice or downstream glacis for dissipation of energy of the falling jet.

- **baffle pier**
an obstruction set in the path of high velocity water to break up the stream, dissipate energy, and prevent scour.

- **baffle platform**
the platform between the toe of the glacis and the baffle wall provided at a level at which the jump would normally form.

- **baffle type scrubber**
a wet scrubber that varies the direction of gas flow by baffles, often used as an entrainment separator and then called a baffle mist eliminator. The pressure drop is low, 2.5 to 7.5 cm water gauge, but many types do not collect particles smaller than 10 microns.

- **baffle type water seal**
a water seal made effective by constructing a baffle or step in the outlet invert.

- **baffle wall**
a cross wall set of vanes or guides, a grid, grating or similar device placed in a conduit to check eddy currents below them, and to effect a more uniform distribution of velocities and to dissipate energy.

- **baffled dryer**
a dryer in which heated air is directed through grain by baffles.

- **baffled outlet pipe drop**
pipe drop having a baffle wall at its outlet.

▪ **baffled side weir**
a weir alongside a sewer, over which the sewage is encouraged to flow by a *baffle* across the main sewer. The baffle can be a vertical, square plate having its bottom edge level with the crest of the side weir, so that the flow over the weir increases as the sewage level rises in the main sewer.

▪ **baffling wind**
wind which is frequently shifting, light airs.

▪ **bag darn**
a low fixed dam made of bags partly filled with soil or concrete.

▪ **bagacello**
fine pith particles of bagasse used to aid mud filters.

▪ **bail plug**
in strainer or slotted pipe tube wells a length of plain pipe is allowed with a cap and a lifting hook to close the bottom end. This cap is known as BAIL PLUG.

▪ **bailer**
an implement, consisting of a section of pipe with a valve in the bottom and a bail at the top, used for removing drillings from the well.

▪ **bailer test**
a test of the rate of depression conducted while boring to estimate the probable yield from any water-bearing stratum.

▪ **bait spreader**
a device which consists of a large, rotating disc onto which poisoned bait is dispensed from a hopper. The disc distributes the bait seven to ten yards from the spreader.

▪ **bajada (bahada)**
coalescing adjacent alluvial fans; a bajada is almost a continuous slope of alluvium along the side of a mountain.

▪ **baked soil**
soil which has a hard, or even glazed surface due to a wet condition followed by an intense drying action of the sun.

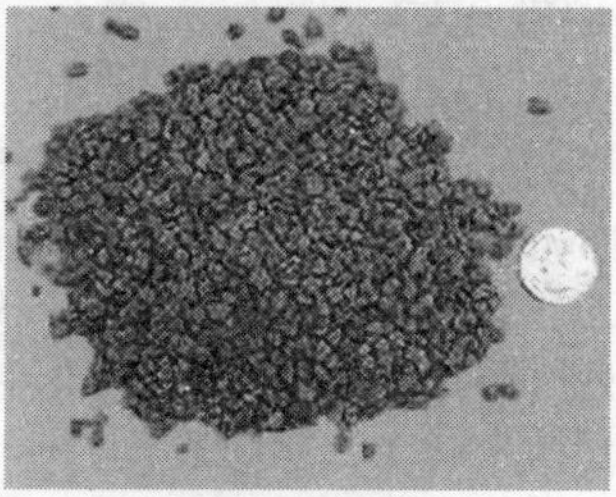

▪ **balanced load**
connection of 120-volt electrical loads at the service entrance panel so that the amperage loads between both ungrounded service conduc-

tors and the grounded service conductor are the same.

- **balanced needle valve**
an improved design valve without Bull Ring or the annular external ring around the needle. The needle is moved by water pressure from the outlet conduit, which acts on interior chambers in the valve. The movement is controlled by a hand wheel installed above the valve, with the motion transmitted through shifting and gearing to a positive positioning device located inside the valve.

- **balanced solution**
a solution of two or more salts, in such proportions that the toxic effects of the individual salts are naturally eliminated.

- **balanced valve**
a valve with both sides of the closing mechanisms in contact with water, and which opens and closes due to the difference of water pressure, or water pressure plus other pressure, exerted upon the sides of the closing mechanism.

- **balanced vane pump**
a kind of hydraulic pump used for preventing unnecessary side-loading of the rotor shaft and bearings.

- **balancing chamber**
a hydraulic chamber in a centrifugal pump to prevent the shaft from bending under the weight and throat of rotor assembly and the water inside it is when pump is in action.

- **balancing coil gauge**
an indicating device (fuel, oil pressure, engine temperature) that contains a pair of coils in the instrument panel unit.

- **balancing depth**
depth of canal cross-section such that the quantity of excavation is just equal to the earthwork required for banks.

- **balancing method**
a means used to find the approximate location of the centre of gravity of track type tractors or other materials.

- **balancing tank**
a reservoir subsidiary to the main reservoir for storing excess river water which is utilised during periods of short supply.

- **bale breaker**
a machine which is used to disentangle, spread and fluff up the cotton fibres after the cotton has been removed from

the bale and before the cotton is spun into thread.

- **bale buster**
a self-powered machine used to break up and shake out tightly baled hay or straw before being fed.

- **bale density**
the weight per unit volume of hay or any other product which can be baled.

- **bale loader**
an elevator which is attached to a truck or tractor for the purpose of lifting bales of hay or straw from the ground to the bed of the truck or wagon.

- **bale press**
a machine used to compress cotton after ginning to bales for shipping.

- **baled**
a product that has been compressed into a large bundle, tied with heavy string or wire and sometimes protected by a wrapping.

- **baled hay**
hay which has been compressed in large bundles, tied with heavy string or wire and sometimes covered with a protective covering.

- **balefill**
controlled tipping by high density baling.

- **baler**
a hay baler in which the pickup cylinder spring teeth pass the hay to an auger or other cross-feed device which in turn passes it to the feeding arms, which feed the hay into the baler chamber.

- **ball and socket joint**
a joint in which a ball or spherical object is placed within a socket recessed to fit it, thus permitting free motion in any direction within certain limits.

- **ball check valve**
a valve consisting of a ball and seat.

- **ball nozzle**
a nozzle, often used for sprinkling, which has a cup at the end of the nozzle and a ball within the cup, the jet issuing from the nozzle striking the ball and being deflected sidewise in all directions.

- **ball planting**
method of planting young plants in which plants are taken from the ground with a compact ball or lump of earth round their roots to provide

nourishment and help their growth in a new location.

- **ball valve**
a device to close a pipe by rotating a sphere in which there is a hole having a diameter equal to that of the pipe, e.g., used at the entrance to a turbine casing.

- **ballast**
crushed stone or broken brickbats.

- **balling up**
accumulation of soft, sticky shale or clay around the bit and drill collar.

- **bamboo rake**
a light rake with teeth made of bamboo pegs. It is used for raking leaves, cleaning beds, etc.

- **band brake**
a curved friction-piece to press against a wheel to stop its turning around.

- **band cutter**
an attachment on the feeder of a threshing machine which cuts the bands on the bundle before delivering the grain to the cylinder.

- **band cutter knives**
the cutting tool of the band cutter.

- **band dryer**
a dryer which dries sludge cake by hot air after pulverisation.

- **band mill**
a saw mill which comprises a heavy cast iron frame and column carrying an upper and lower band wheel.

- **band placement**
the placement of fertilisers in the soil in continuous band or narrow ribbons in between the plant rows or around each plant.

- **band saw**
an endless band of steel with saw teeth upon one edge, passing over and driven by two wheels.

- **band screen**
a slopping conveyor belt of wire mesh (e.g. 6 mm holes) through which sewage or water flows, leaving the larger solids on the screen.

- **band seeding**
placing the seed in rows directly above, but not in contact with a band of fertiliser.

- **band shoe**
a metal fitting used for fastening and tightening the rods or bands which are wrapped around wood stave pipe.

- **band wheel**
  a pulley, over which a belt runs.
- **bandal**
  a fascine consisting of framework with bamboos, etc., driven into the river bed, say 2 feet (61 cm.) apart with horizontal ties and supported by struts placed at every 4 feet (121 cm.) and a screen or mat of split bamboos.
- **bandalling**
  a system of river training in India by means of bandals designed to confine the low eater flow in a single channel with a view to maintaining a suitable depth for navigation, as also for having greater head for irrigation.
- **banded over seed**
  placement in a band over the seed row.
- **banded steel pipe**
  a steel pipe whose strength has been increased by use of bands which are shrunk around the shell.
- **banding fertiliser**
  the method of placement of commercial fertiliser in concentrated strips.
- **bank**
  a place which has not been turned over by the plough.
- **bank erosion**
  land or soil losses of the banks of rivers, lakes, ditches or canals resulting from currents, wave cutting, undermining, caving, rilling, or gullying of slopes.
- **bank full**
  a term used in irrigation to describe the condition of a ditch which is completely filled with water but not overflowing.
- **bank measure**
  measurement of soil or rock based on the quantity material in place on the cut.
- **bank paving**
  the part of revetment extending from the mattress to the top of banks.
- **bank protection**
  a kind of engineering work which aims at the protection of banks of a river, or slopes of embankments along it, from erosion by the current of flow.
- **bank protector**
  any device or material used to stabilise the sides of a river or ditch.
- **bank revetment**
  a type of bank protection which covers continuously

the entire slope of a bank or an embankment, including the portions extending far into the river bed, to protect the bank from erosion.

- **bank run gravel**
normal product of a gravel bank, including pebbles and sand in varying proportions.

- **bank sluice**
an opening controlled by shutters or gates in the marginal embankment which are opened during high floods.

- **bank storage**
water absorbed and stored in the bed and banks of a stream, lake or reservoir, and returned in whole or in part as level of the surface of the water body falls.

- **bank strengthening**
a general term which includes widening, raising and straightening of banks of a channel; and also forming by silting tanks, banks of a new channel passing through low ground.

- **bankful**
the stage or gauge height attained by a river or stream when flowing at capacity above which banks are overflowed.

- **bankful discharge**
discharge corresponding to the state at which the river berms are just about the submerged.

- **bankful stage**
the stage at which a stream just overflows its natural banks.

- **bankside storage**
raw water storage in a reservoir on the river bank.

- **banquette**
a berm or bench formed , on the landside of a levee, for the purpose of increasing the width of the lower portion thereof, to prevent the plane of saturation from cutting the side of the levee and causing it to slough; also to provide a break in the surface drainage.

- **bar chart**
the graphical representation of frequencies or magnitudes by bars drawn with lengths proportional to the frequencies or magnitude concerned.

- **bar draft**
  the depth of water over a bar as measured at low tide.

- **bar graph**
  is a diagram consisting of a series of bars or columns proportional in length to the quantities they represent.

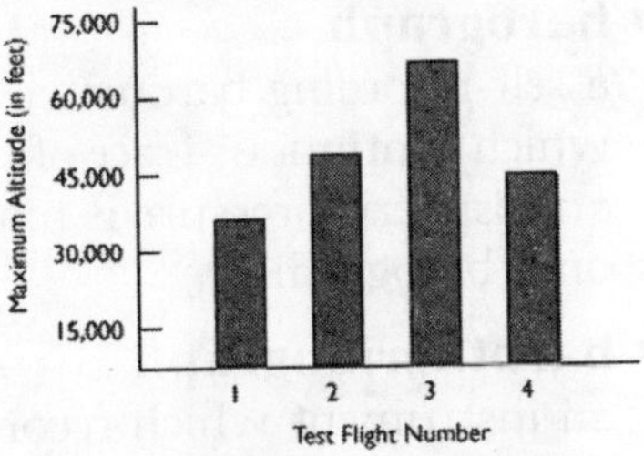

- **bar hanger**
  in house wiring, a device for supporting outlet boxes. The hanger is attached to joists.

- **bar port**
  a harbour that can be entered only when the tide rises sufficiently to permit passage of vessels over a bar.

- **bar rack**
  a screen composed of parallel bars, either vertical or inclined, placed in a waterway to catch debris, and from which the screenings may be raked.

- **bar roller**
  a device which is used for crushing clods and compacting soil.

- **bar spade**
  a spade having a cutting blade welded across three parallel bars (similar to a garden fork) for trenching in wet, sticky soil.

- **bar tilting adjustment**
  an adjustment on the cutter of a mover which changes the angle of the cutter bar in relation to the ground surface over which it passes.

- **barbed wire**
  strands of steel wire twisted together, with sharp pointed barbs or hooks clenched or interwoven between them at uniform distances.

- **barbel region**
  the middle reaches of a river having sandy or silty bed and deep waters; the principal sedentary fish barbel live and others like carp, pike and perch also live.

- **barbier morgan index**
  an index of the chemical fertility of the soil.
- **barff's process**
  a method of protecting iron from rusting by oxidising it with superheated steam.
- **barley fork**
  a large long handled fork which has a guard at the handle end of the tines and is used to pick up loose short stemmed straw for transporting.
- **barley reel**
  a seed separating machine which consists bar of a rotating, corrugated, woven screen.
- **barn cleaner**
  a power operated conveyor type of mechanical or chain device which runs in the manure gutter for removal of manure and litter.
- **barn drying**
  the removal of moisture from hay and other forages within the barn using unheated forced air.
- **barn hay drier**
  a device used for drying hay in the barn using forced air, which is usually unheated, and a duct system for distribution of this air.
- **barnyard**
  the enclosed ground around a barn wherein animals are allowed to exercise.
- **barogram**
  the continuous record of atmospheric pressure made by a barograph.
- **barograph**
  a self-recording barometer, in which continuous trace of the atmospheric pressure is made on a barogram.
- **barothermograph**
  an instrument which records simultaneous .barometric pressure and temperature on the same chart.
- **barrage**
  a barrier, provided with a series of gates, across the river to regulate water surface level and flow upstream and to divert water supplies into a canal; distinguished from a weir in that it is gated over its entire length and may not Tmve a raised still
- **barren lands**
  flat open lands with vegetation limited to small trees and sparse shrubs.
- **barrier**
  a low dam or sill in a small stream usually in a moun-

tainous region, built to retard the downward cutting of the stream.

- **barrier bar**
  a ridge of sand or gravel lying more or less parallel with but some distance from the mainland so as to separate a bay or river mouth from a larger body of water.

- **barshy soil**
  soil composed of coarser angular rock fragments with fine soil.

- **barysphere**
  the central or deep interior portions of the earth presumably composed of heavy metals or minerals.

- **basal dose manure or fertiliser**
  applied to soil before or along with sowing or transplantation of crops. Organic manures, phosphatic and potassic fertiliser are always applied as basal dose. Regarding nitrogen, a part or whole of the requirement is applied as a basal dose.

- **basal dressing (base dressing)**
  fertiliser (usually P and K) applied before establishment of a crop or orchard to provide for its needs over a season or period of years.

- **base course**
  a layer of specified or selected material of planned thickness constructed on the sub grade or sub base for the purpose of serving one or more functions such as distributing load, providing drainage, minimising frost action, etc.

- **base data analysis**
  comprises the computation and synthesis of available data for delineating homogeneous areas and for determining research priorities and strategies.

- **base distance**
  the distance from the line of sight to the grade line.

- **base exchange capacity**
  the amount of base held on a clay under specified conditions, e.g., of pH.

- **base exchange in soil**
  the physico-chemical process whereby one species of ions adsorbed on soil particles is replaced by another species.

- **base flow depletion curve**
  the base flow of a stream at a particular point diminishes at a characteristic rate until

supplemented by precipitation or snowmelt. This characteristic rate is defined by the base flow depletion curve.

- **base flow, basic flow**
the sustained or dry weather flow of streams resulting from the outflow of permanent or perched ground water, and from the draining of large lakes and swamps.

- **base level**
the theoretical limit of land erosion.

- **base level of erosion**
the lowest level to which a stream can erode its bed.

- **base line**
a survey line established with more than usual care, to which surveys are referred for co-ordination and correlation.

- **base load**
the minimum load over a given period of time.

- **base map**
a map having sufficient points of reference, such as state, county, or township lines, and other selected physical and cultural features, to allow the plotting of other special data.

- **base of the notch**
the bottom horizontal surface of the notch.

- **base rich**
applied to soils containing a relatively large amount of free basic ions such as calcium or magnesium.

- **base saturation percentage**
the extent to which the adsorption complex of a soil is saturated with exchangeable cations other than hydrogen and aluminium. It is expressed as a percentage of the total cation exchange capacity.

- **base saturation ratio**
ratio of the base content to the humus content. The affinity of humus for the base decrease as the ratio increases.

- **base sedimentation survey**
sedimentation survey normally made prior to the completion of a new dam and before the storage of water has begun.

- **base year**
a year taken as the standard (initial year) for an economic index.

- **basic holding**
a holding which though uneconomic in the sense of being unable to provide a reasonable standard of living to

the cultivator, may not be inefficient for the purpose of agricultural operations.

- **basic intake rate**
rate at which water will enter the soil when after initial wetting of the soil the rate become essentially constant; which is equal to saturated hydraulic conductivity mm/hr.

- **basic stage flood**
an arbitrary selected rate of flow of a stream used as the lower limit or base for selecting floods for analysis; sometimes taken as the minimum annual flood.

- **basic type double tier formwork**
formwork which consists of panels so designed that the lower ones serve as support for the upper panels; their vertical displacement proceeds by a successive transfer of the lower panel to the next upper panel position, i.e., the upper and lower panels are of the same dimensions and interchangeable.

- **basic type single tier formwork**
formwork consisting of panels with studs of a length equal to the height of sheathing.

- **basin cultivation**
the growing of crops in small hollows or basins, often formed by raising low ridges of soil around a plot of land.

- **basin irrigation**
the method of applying irrigation water to crops, wherein small plots are surrounded by borders, and filled with water.

- **basin lister**
a double plough which leaves the field in a series of alternate ridges and furrows. The operation helps in detention storage of rainwater.

- **basin listing**
tillage that dams lister furrows at regular intervals to create small basins.

- **basin lock**
lock built where a navigation canal passes through a basin.

- **basin method**
a method of artificial recharge in which water is impounded in a series of basins that are formed by a network of artificial dikes or levees, generally using to full advantage the surface contours.

basin method of irrigation
a method of irrigating orchards by which each tree or a group of trees is surrounded by a border, to form a pool or a small basin when water is applied. The water is usually turned direct from the supply ditch into the basin but may sometimes be turned from one basin into another.

basin recharge
that portion of the storm rainfall that is required to satisfy the demands of interception, depression storage, and soil moisture. It is the difference between rainfall and runoff and is often referred to as loss.

basket dam
a sack dam composed of boulders held in a woven wire sack and piled up to form a barrier or jetty.

basket planting
planting young trees from or with the loosely woven baskets in which they have been sown or transplanted.

bastard file
a file intermediate between the coarsest and the second cut.

bastard trenching
a laborious system of bringing subsoil to the surface in gardens in which two-ft. ditches are dug. The earth from the first is placed aside and the earth from the second is placed in it. Next, a layer of manure is placed in the bottom of the ditch along with a layer of earth from a third trench and so on.

bastard wheel
occasionally used to denote a bevel wheel whose amount of bevel is so slight that it approaches very nearly to a right angle or to a spur wheel. Also one accommodated to an existing wheel, though not absolutely correct.

batch
a quantity of a harvested crop or any other material which is put into a bin or container on a repetitive basis specifically for treatment, such as drying. The container may be utilised for storage of material after drying in contrast to continuous operation. The quantity of anything produced at one operation.

batch bin
a structure used for drying small batches.

- **batch meter**
  a meter which weighs or measures batches of aggregate and, usually, installed with dial scales for visual control.

- **batch mixer**
  a concrete mixer in which all the ingredients for a batch of concrete are placed at the same time, and no additional material (except water) added during mixing period or until the batch is completely discharged from the mixer.

- **batch peeler**
  a machine which is used to peel a fixed quantity of fruit or vegetable at one time, as contrasted to a continuous peeler.

- **batch treatment**
  any treatment in which the process is completed and the products are discharged before more raw material can be taken in.

- **batchers**
  mechanical equipment to measure out or to weigh batches of the different ingredients of a concrete mix.

- **batching plant**
  a plant that has bins with separate compartments for fine aggregate and separate compartments for each size of coarse aggregate.

- **bathymeter**
  an instrument used for determining ocean depths.

- **bathythermograph**
  an instrument that measures ocean temperature at depth.

- **batter**
  the slope given to a wall or an earthwork.

- **battery**
  1. two or more chemical cells connected together that produce electricity by chemical reactions.
  2. a device that stores electrical energy using electrochemical cells. The physical structure of battery does not permit the intermixing of chemicals stored in it.

- **battery ignition system**
  an electrics spark type ignition system using a battery as a source of current for the primary circuit.

- **battery of siphons**
  a number of siphons installed close to each other.

- **battery of wells**
  two or more wells that are connected to common pump section manifold. It is used to obtain a greater flow of water than is available from a single well.

▪ **bayou**
a distributary channel for flood water, common to streams that flow through alluvial valley land. (2) Channels through swamps and marsh lands, including those that afford -passage for tidal water. (3) Abandoned channels of the parent streams. *Also see* oxbow lake.

▪ **Bazin roughness coefficient**
the roughness coefficient in the formula proposed by Bazin for determining the discharge coefficient in the Chezy formula.

▪ **bc soil**
soil with B and C horizons with little or no A horizon.

▪ **bca horizon**
a horizon enriched in calcium.

▪ **bear trap dam**
a movable dam consisting of a masonry sill, masonry piers, and a series of bear-trap gates installed on the sill between the piers.

▪ **bear trap shutter (gate)**
a type of automatic shutter consisting essentially of two leaves, hinged and sealed at the base. The two leaves also have a sliding hinge or seal at their juncture.

▪ **bearing surface**
the area of the surface upon which a shaft rotates.

▪ **beater**
any mechanical device that pounds, stirs, loosens, as a revolving blade mechanism in a threshing machine directing the straw to or form the cylinder or as found at the rear of manure spreaders.

▪ **beaufort scale**
a numerical scale used by seamen showing the velocity of the wind in statute miles per hour with the corresponding velocity in nautical miles per hour and the terms applied to the various limits of velocity. In customary usage the designation is applied to the set of terms which describe velocity ranges of the wind.

▪ **bebout wicket dam**
a movable dam similar to a Chanoine wicket dam with exceptions: The legs of the horse, instead of being rigid, have a knee joint, and the prop

consists of two parallel members that are hinged to the base slab instead of a single member sliding in a groove.

- **beccari system**
a composting system which is unusual in being partly anaerobic, although it is not foul-smelling.

- **bed**
narrow flat-topped ridge on which crops are grown with a furrow on each side to facilitate irrigation and for drainage of excess water.

- **bed building stage of river**
the river stage corresponding to the maximum of the sediment runoff curve.

- **bed groin**
a groin built slowly up from the bed, with the effect of progressive and even reclamation over a large area.

- **bed levels**
levels (designed or average of the cross-section) of the bed of a canal at any point along the centre line.

- **bed load**
sand, silt, gravel, or soil and rock detritus carried by a stream on or immediately above its bed. The particles of this material have a density or grain size such as to preclude movement far above or for a long distance out of contact with the streambed under natural conditions of flow.

- **bed load feeder**
an apparatus for artificially feeding bed load in hydraulic model experiments.

- **bed load function**
the rates at which various discharges will transport the different grain sizes of the bed material in a given channel.

- **bed load rate**
the rate of bed load movement, expressed in weight or volume of bed load transported over the entire width of the river bed, or per unit width of river bed and per unit of time.

- **bed load sampler**
an instrument which directly measures the rate at which sediment moves in the layer just above the bed of a stream.

- **bed material**
the sediment mixture of which the moving bed is composed.

- **bed material load**
that part of the sediment load which consists of grain sizes represented in the bed.

- **bed material sampler**
an instrument to sample the bed material, whether in motion or at rest.

- **bed method**
involves levelling of land and preparing small beds surrounded by bunds.

- **bed pitching**
pitching on the bed of the earthen canal sections, upstream and downstream sides (usually beyond the transitions) of canal structures for protection against light scouring action of water.

- **bed planting**
planting on elevated level beds, which are separated by furrows.

- **bed plate**
steel plate fixed at the bed of the controlling section of an outlet so that the bottom of the section may not be tampered with.

- **bed ripples**
undulating ridges and furrows or crests and troughs formed by the action of flow on the bed of a channel. Specifically 'ripple marks'.

- **bed slope**
the inclination or gradient of bed from a horizontal line, usually expressed as a ratio, such as 1 in 100 (one unit rise in 100 units of horizontal distance), or in a decimal fraction (0.01).

- **bed up**
to build up beds or ridges with a tillage implement.

- **bed waves**
wave of sediment formed by the action of flow on the bed of a channel.

▪ **bed width**
in rectangular or trapezoidal section, the distance between the extreme points at the bottom of a cross-section.

▪ **bedding**
a method of surface drainage of narrow-width plough land in which the dead furrows (depth varying from 0.5 to 1.5 ft. or 15.24 to 45.72 cm) run parallel to the prevailing land slope.

▪ **bedding land**
ploughing or otherwise raising the surface of a field into a series of parallel beds separated by dead furrows or shallow drains. It is useful in improving field drainage.

▪ **bedding of soils**
surface drainage through the provision of shallow drains framed by cultivating the land into beds (by ploughing in narrow strips or lands) the dead furrow is left open to serve as a watercourse.

▪ **bedding plane (or rock)**
refers to the plane of junction between different beds or layers of rock.

▪ **beet harvester**
any machine which is used for harvesting beets mechanically.

▪ **beet lifter**
a device consisting primarily of two prongs that pass through the soil and straddle the row to partly lift sugar beets out of the ground.

▪ **beheaded stream**
a stream whose original headwater drainage area, separated from it by another stream which was more active and had greater erosive power, now drains into the other stream.

▪ **bell crank**
a triangular frame oscillating about one angle as pivot, used to convert horizontal reciprocal motion into vertical, or *vice versa*.

▪ **bell mouth**
the streamlined shape of the upstream end or edge of an opening constructed to ensure entry of flow into the off taking closed conduit or channel.

▪ **bell mouth chamber**
a chamber at a junction between two or more large sewers, in which the side of the outgoing sewer forms a tangent to the incoming sewers.

▪ **bell mouthed orifice**
an orifice with a short tube on the outside, the tube flaring or

increasing in diameter away from the opening in the orifice plate.

- **bell ringing transformer**
  a small transformer to furnish current at proper voltage for ringing electric bells.
- **bell socket**
  a connecting piece for two different diameter pipes used in a compound well. The socket is screwed to both the pipes at their junction.
- **bellows**
  a device, usually metal, that can lengthen or shorten much like an accordion.
- **bellows duster**
  a portable duster which consists of a bellow, a discharge chamber, and a pipe.
- **belt**
  a broad strip of land characterised by one dominant feature, such as a crop or climatic type.
- **belt clamp**
  device, a stretching frame, the two ends of which are coupled by screwed bars.
- **belt cleaner**
  a device at the discharge end of a belt conveyor to prevent a portion of the mix adhering to the belt.
- **belt conveyor**
  a conveyor with an endless belt for transporting concrete to the job site.
- **belt distributor**
  a system of conveying seed cotton within a gin on 8 to 12 in. belts, which have 4 in. spikes, fixed at regular intervals.
- **belt dressing**
  a material applied to flat belts to prevent slipping on the pulleys.
- **belt lacing**
  a material (usually leather) used to fasten two ends of a belt together. Metal laces are used on belts transmitting small amounts of power.
- **belt of no erosion**
  the upper part of a hill-slope, extending from the crest for some distance down-slope, on which the surface accumulation of water from precipitation is insufficient to cause erosion.
- **belt of soil water**
  the part of the zone of aeration which, consists of soil and other materials that lie near enough to the surface to dis-

charge water into the atmosphere in perceptible quantities by the transpiration of plants or by evaporation from the soil.

- **belt of water table fluctuation**
that part of the lithosphere which, because of the fluctuation of the water table, lies in the zone of saturation part of the time, and a part of the time in the overlying zone of aeration.

- **belt power**
the maximum sustained power, measured at the dynamometer after belt transmission, with the controls set as specified by the tractor manufacturer for pulley work.

- **belt shifter**
a device for placing a belt on a pulley while the latter is in motion.

- **belt tests**
tests designed to measure power, fuel efficiency govern control and horsepower ratings of engines transmitting their power through a belt.

- **belt transect**
a narrow strip of land taken as a sample of an area in studying the vegetation.

- **belt type seed metering device**
a device which utilises a belt to effect metering of seeds in a planter. The belt commonly has holes or depressions along its face, which convey seeds from a hopper to the discharge chute and thereby meter the quantity of seed by their size and spacing.

- **bench**
a narrow terrace-like land form, characterised by a flat surface with a steep slope at the back. Benches may be erosional in origin, the most common being produced by Jakes or rivers that have experienced a change in base level.

- **bench border method**
a method of surface flooding in which water flows over long narrow strips separated by low earth a banks located on cross-slope benches having a slight slope in the direction of strip.

- **bench flume**
a flume supported on a shelf or a bench cut in a hillside or built around mountain slopes or set on ground.

- **bench gravel**
gravel beds which occur on the sides of the valleys above

the present stream bottoms, representing parts of the bed of the stream when it was at a higher level.

- **bench irrigation**
a method of irrigation by flooding adapted to fairly uniform, moderate slopes with deep soils.

- **bench shear**
a shear mounted upon a bench and used for the cutting of sheet metal.

- **bench terrace**
a terrace designed to facilitate soil and conservation, erosion control, and cropping on steep slopes by the development of level or flat gradient strips normal to the slope used for agricultural purposes.

- **bench terrace water harvesting system**
the system uses most of the principles of runoff farming to conserve and utilise runoff for grain production.

- **benching**
ledges shaped like steps or terrace formed below beds of canals and under the seats of banks in high filling for proper bounding of earthwork with the natural ground.

- **benchmark**
a permanent point on a monument or permanent structure, whose elevation above a datum is known or fixed by precise levelling and which is used as a point of reference in the determination of other elevations.

- **bend**
a change in direction of, for example, a river.

- **bending stress**
the stress produced in a member by a bending moment, consisting of a combination of tension, compression shear.

- **bendix drive**
a type of cranking motor drive which meshes with the flywheel teeth for cranking. It uncouples automatically as engine speed increases.

- **benefit cost analysis**
evaluation, quantitatively and qualitatively, of the positive and negative effects of a project.

- **bentometer, bendometer**
a calibrated bend or elbow with suitable upstream and downstream straight lengths for measurement of flows.

- **bentonite**
  a generalised term which refers to a rather heterogeneous substance, composed mainly of the clay mineral mont-morillonite and fragments of other minerals, such as feldspar, gypsum, calcium carbonate, quartz and traces of other minerals.

- **bentonite grouting**
  a process in which bentonite, with or without other ingredients, in the form of suspension, is injected under pressure into a previous medium to make it watertight.

- **bentzel tube**
  an instrument designed by Bentzel for measuring velocity of fluids.

- **berkefeld filter**
  a domestic drinking water filter consisting of a metal cylinder containing compressed diatomite through which the water flows. It can be cleaned by backwashing.

- **berm**
  a shelf that breaks the continuity of slope of a bank or cutting.

- **berm ditch**
  a ditch constructed along a berm of ground to carry away surface runoff.

- **berm trimming**
  cutting away the overgrown, overhanging portion of the berm to restore it to its final regime shape.

- **berming**
  the deposition of material on the sides of a channel forming a berm.

- **bermuda grass cutter bar**
  a cutter bar for a mowing machine which has twice as many guards in the same length as a regular bar and which has ledger plates on each guard.

- **beta error**
  the acceptance of a null or other hypothesis when it is really false.

- **beta factor**
  the ratio of the air saturation value in an activated sludge to that in pure water at the same temperature and pressure.

- **beta particle**
  an electron, either positive or negative. Positive beta particles are called positrons. Negative beta particles are called negatrons.

- **betonitic clay**
  a clay with a high content of the mineral montmorillonite,

usually charac-terised by high swelling on wetting.

- **bevel wheel**
a toothed wheel shaped like a conical frustum, used in pairs to transmit motion.

- **b-horizon**
soil horizon generally beneath A horizon in which clay, iron or aluminium with accessory organic matter have accumulated and blocky or prismatic structure had developed. In distinct soil profiles, it is roughly equivalent to subsoil.

- **bi-directional triode thyristor (TRIAC)**
a solid-state electronic switch which allows current to flow in both directions.

- **biennial weeds**
weeds having a life of two years only.

- **bi-flow filter**
a sand filter with upward flow in the lower half of the bed and downward flow in the upper half.

- **bifurcation**
a forking, or division into two branches; the point of forking; a branch.

- **bifurcation gate**
a gate in a structure which divides the flow between two conduits.

- **big valley sweep**
one of many designs of cultivator shovels.

- **bill**
a part of the knotter assembly found on grain binders and automatic tie-hay balers.

- **bill hook**
a metal hook resembling a bird's neck, used to grasp string in a twine knotter, which is a part of the knotter head on a machine for tying string; for example, a binder.

- **billabong**
a blind channel leading out from a river; a backwater.

- **billet**
a steel slab for distributing a load (such as a column load) to a grillage or footing.

- **billow**
usually a great wave or surge of water; any wave.

- **bimetal**
referring to the thermostatic bimetal element made up of two different metals with different heat expansion rates.

- **bimodal size distribution**
a particle size distribution in which most of the grains are concentrated around two sizes.

- **bin**
a box, crib, or enclosure which is used for storage, particularly the storage of grains, machinery parts, etc.

- **bin drier**
a large receptacle for drying grains or other materials, usually fitted with either hot or cold air circulating means.

- **binary sorter**
in a mechanical sorting plant, the usual device that divides the material into two parts, one with and one without the constituent being sorted.

- **binder**
the fine soil (clay) fraction inherent in or added to sands and gravels for cohesion; also admixture for stabilising.

- **binder course**
an intermediate course between the base and the surface course. It is composed usually of an open-graded mix of stone and sand aggregates, with little or no filler, and bitumen or cutback.

- **binder twine**
a coarse string or cord usually made of sisal or hemp which is used in harvesting machinery.

- **bindweed sweep**
one of many designs of cultivator shovels.

- **bioaccumulation**
build-up of some specific organic or inorganic compound within a specific tissue of a given organisms usually applied to certain heavy metals, pesticides or metabolities.

- **bio-aeration**
a modification of the activated sludge process in which the sewage and sludge are agitated and aerated by mechanical means, such as paddle wheels, turbines, or ejectors.

- **bio-assay**
a trial of a chemical or a pollutant, made by testing how much of it can be tolerated by a living organism.

- **bioclimatic law**
the maturity of a given crop may vary by 4 days for each 1° latitude, 5° longitude and 400' altitude.

- **biodigested slurry**
the slurry that remains in the digestion chamber after production of biogas.

- **bio-flocculation**
  the aeration of settled sewage with activated sludge for 1 h, followed by treatment of the effluent on trickling filters. The activated sludge is re-aerated.

- **biogas**
  a mixture of gases containing methane, carbon dioxide, hydrogen and traces of few other gases produced by the anaerobic fermentation of easily decomposable cellulosic materials in presence of methane forming bacteria.

- **biogeochemical cycle**
  the cycling of such chemical elements as nitrogen, carbon, and sulphur from the environment into organic substances and back into the environment.

- **biogeochemical cycling**
  the cycling of chemical constituents through a biologic system.

- **biogeography**
  that part of geography which deals with aspects of animal and plant life particularly their distribution.

- **biological film**
  a film of slime that forms in a micro-strainer, rotating biological contactor, slow sand filter, trickling filter, etc.

- **biological filter**
  a trickling filter.

- **biological indicator (or index) of pollution**
  the presence and frequency of a plant, animal, microbe or other form of life can give an idea of the level of pollution in its environment. But an organism that indicates pollution may provide no indication about poisons. Various classifications of organisms indicating levels of organic pollution have been produced, e.g., the seprobic classification, the Trent biotic index.

- **biological interchange**
  the interchange of elements between organic and inorganic states in a soil or other substrate through the agency of biological activity.

- **biological monitoring**
  repeatedly surveying plants and animals at a site to determine their abundance, at least roughly, and noting the physical, chemical and biological changes with time, as in a transect.

▪ **biological weathering**
refers to physical and chemical weathering assisted by biological agencies.

▪ **biological weed control**
a method of weed control consisting of either shading of water, or introduction of suitable type of fish or mammals who live on specific weeds.

▪ **biological yield**
refers to total dry matter produced by a plant or of a crop in unit land area.

▪ **biomass**
any organic matter which is available on a renewable basis, including agricultural crops and agricultural wastes and residues, wood and wood wastes and residues, animal wastes, mineral wastes and aquatic plants, usually expressed as dry weight per unit area.

▪ **biomass density**
weight of living material expressed as a dry weight, in all or part of an organism, population, or community, expressed as weight per unit area.

▪ **biometer**
a method of assessing the amount of life by measuring the respiration, e.g., in a soil sample.

▪ **bionomics**
study of the relation of an organism or population of organisms to its environment.

▪ **biopedology**
that part of soil science which studies the living beings (plant and animal) of the soil and their influence on it.

▪ **biophysics**
the study of phenomena of living organisms by physical methods; the study of physical phenomena exhibited by living organisms or parts thereof.

▪ **biosequence**
a sequence of soils whose properties are functionally related to differences in organisms, as a soil forming factor.

▪ **biosorption**
a contact stabilisation method used in Texas. The surplus sludge is disposed off in lagoons, which function well in the sunny climate.

▪ **biota**
the living parts of an ecosystem.

▪ **biotape**
an area where the principal habitat conditions and the liv-

ing forms adapted to the conditions are uniform.

- **biotechnology**
  the application of biological organisms, systems or processes to manufacturing and service industries.
- **biotic adaptation**
  changes in form or physiology, presumed to have arisen as a result of competition with other plants.
- **biotic index**
  a numerical classification of the life in a river, which indicates the quality of the water, as in the Trent biotic index.
- **biotic potential**
  the inherent power of an organism to multiply over a given period of time in the absence of control factors.
- **biotope**
  a relatively small area where the environment is uniform.
- **bipolar transistor**
  a type of transistor that consists of three alternating layers of p- and n-type semiconductor material.
- **bird scatter**
  mechanical device used for scaring away birds.
- **bird's mouth**
  mark made in a rafter to fit the plate.
- **birth of a soil**
  a soil is born as soon as a portion of the parent rock is exposed to the action of the atmosphere and living organisms.
- **bit**
  the portion of a drill which contacts the rock and disintegrates it.
- **bit stop**
  an attachment to a bit which limits drilling or boring to a given depth.
- **bitumen**
  those substances which, extracted from the soil, are insoluble in water and soluble in the organic solvents.
- **bitumen cement concrete lining**
  a type of lining in which concrete is a mixture of bitumen cement and aggregates graded from fillers and fine sand to coarse gravel.
- **bitumen grouted construction**
  a layer of coarse stone of more or less uniform size, grouted with bitumen.

- **bituminous macadam**
an open graded mix of stone and sand aggregates, filler in small amount, and bitumen or cutback.

- **black locust**
a decay resistant hardwood tree especially good for fence posts.

- **blade**
a cutting tool as the blade of an adze or the blade of a bulldozer.

- **blade harrow (bakhar)**
a harrow which consists of one or more blades attached to the beam or frame used for shallow working of the soil with minimum of soil inversion.

- **blank**
a piece of timber cut to a specialised size and shape from which the finished article is made.

- **blank bolts**
rough forged bolts previous to threading.

- **blank flange**
a pipe flange that is not drilled but which is otherwise complete.

- **blanket**
a cover of one type (pervious or impervious) material by a layer of another type of material.

- **blanket application**
the application of spray or dust over an entire area rather than on rows or beds.

- **blasting**
discharging of an explosive to loosen rock.

- **blasting cap**
a small metal tube loaded with a charge of sensitive explosive.

- **blasting machine**
machine used to generate the current for firing as explosive by electricity.

- **blasting powder**
this is a slow burning low explosive made from saltpetre, sulphur and charcoal. It is seldom used for blasting rock.

- **bleed**
the procedure of removing air from a diesel fuel system.

- **bleeding**
the discharge or freeing of water from freshly placed concrete.

- **blemish**
anything that mars the appearance of wood but is not serious enough to be classed as defect.

blended cement
the impalpable powder obtained by mixing Portland cement with more than one per cent of an admixture other than a retardant, for the purpose of obtaining the normal properties of the Portland cement.

bligh's creep theory
the premise of this theory is that the resistance to percolation along the 'line of creep', the line of contact of the dam and cut-off with the foundations may be less than directly through the foundation materials on account of the difficulty of securing an intimate contact and that the hydraulic gradient is constant throughout the creep length.

blind drain
in an earth dam, a lens or stringer of coarse and more or less permeable material due to poor construction, which may extend from the upstream face inward and afford opportunity for water to pass into and possibly through the dam; if the percolating water carries away sufficient fine material, failure of the dam may result.

blind flange
a pipe flange used to close the end of a pipe.

blind inlet
a porous backfill section over a tile line which is constructed with selected porous materials to prevent sediment from entering the line.

blind level
a level drainage tunnel having a shaft at either end, and acting as inverted siphon.

blinding
blocking by fine material of a slow sand filter or a wire cloth mesh in a filter press, screen, vacuum filter, etc.

block
the portion of a dam between two transverse joints constructed normal to the axis.

blocking hoe
a hoe with a straight shank in which the blade is in direct line with the handle, used for thinning.

blocks
obstructions set in the path of high velocity water, such as piers, on the apron of an overflow dam, weir or drop, to dissipate energy, prevent scour downstream and also to control the position of hydraulic jump.

■ **blow**
eruption of sand and water inside a coffer dam, causing failure of flooring.

■ **blow by**
leakage of compressed air fuel mixture or burned gases from the combustion chamber, past the piston rings, and into the crankcase.

■ **blow off valve**
a valve installed in a low point or depression on a pipe line to allow drainage of the line.

■ **blow pipe**
a tube by which air or gas is blown through a flame for heating, fusing, or melting.

■ **blower duster**
a machine used to apply dust, insecticides, herbicides, etc., using air as a vehicle.

■ **blowing well**
a well from which a current of air issues. Such a current is sometimes fairly strong and noticeable.

■ **board foot**
a unit of measure for lumber consisting of a volume equal to 144 cubic inches.

■ **boiler feed water**
water forced into a boiler to take the place of that which is evaporated in the generation of steam.

■ **boiler scale**
an incrustation varying from a porous, friable crust to a dense, very hard coating deposited on boiler heating surfaces due to precipitation of minerals out of solution from the water used.

■ **boison**
a topographic basin with an inward flowing drainage system that drains to a playa.

■ **bollard**
a steel or concrete post installed near or on the side walls for securing ropes to for mooring purposes and also for retaining vessel entering the lock.

■ **bolometer**
an instrument of great sensitiveness, consisting essentially of a Wheat stone bridge to measure minute quantities of heat energy.

■ **bolster**
a bag of broken stone enclosed in a wire cage used in the construction of an underwater embankment, for assistance in closing of canal breaches or checking erosion, or filling of scours below or adjoining

hydraulic structures in rivers or canals.

- **bolt**
  a pin of iron or brass used to fasten or secure something in place.

- **boost**
  the increase in an engine's intake air pressure produced by a turbo-charger.

- **bottom dead centre (BDC)**
  piston position at the exact bottom of its stroke.

- **bow instrument**
  mechanical drafting instrument used for producing circles and arcs.

- **Boyle's Law**
  physical law describing the relationship between the pressure and volume of a gas at constant temperature.

- **branch circuit**
  electrical circuit from the last over-current protection device to the utilisation equipment.

- **brazing**
  a non-fusion welding process.

- **bread boarding**
  testing and assembling a provisional electronic circuit before constructing a permanent version.

- **bridge rectifier**
  a four-diode network used for changing both halves of an incoming AC sine wave into DC electricity.

- **BTU (British Thermal Unit)**
  a measure of heat energy 1 BTU is the quantity of heat required to raise the temperature of 1 pound of water by 1 degree Fahrenheit.

- **building service conductors**
  electrical conductors that supply electricity from the central distribution point to individual load centres.

- **byconversion**
  a general term describing the conversion of one form of energy into another by using biotic sources like animals, plants or microorganisms. Synthesis of organic compounds from carbon dioxide by plants is bioconversion of solar energy into stored chemical energy.

- **c horizon**
  soil below the b horizon which is important for storing and releasing water to the upper layers of the soil.

▪ ca
a sub horizon symbol indicating an accumulation of carbonates of alkaline earth, commonly calcium.

▪ cable
two or more electrical conductors that are enclosed in a protective outer sheath.

▪ cable ripper
an electrician's hand tool used for slicing the outer cover of a cable.

▪ cable tool drilling
a method of drilling wells by the use of cable tools.

▪ cableway excavator
a cable excavator having generally a long range track cable which is loosened to lower the bucket, and tightened to raise it.

▪ cadelle lava
one of the principal insects which attack grains in storage.

▪ cage
the receptacle in which men and materials are raised or lowered in a shaft by means of winding or hoisting gear.

▪ cage bar
thick, parallel steel bars that act as a *screen* in the *grate* under a horizontal-shaft hammer mill.

▪ cage disintegrator
a mill with a rotating cage having horizontal bars to reduce the size of brittle material.

▪ cage house
a type of poultry house in which all of the fowl are housed in cages either individually or in small groups.

▪ calcareous pan
a hard layer of limey material, more or less impermeable to water, formed below the surface of the soil, and affecting the water supply of plants growing above it.

▪ calcareous soil
a soil containing sufficient calcium carbonate to effervesce when tested with weak (0.1 N) hydrochloric acid.

▪ calcifugous plants
a plant species which does not thrive on a lime soil.

▪ calculated drainage
the water height D (mm) which infiltrates in one year in a given soil.

▪ calibration
the process of evaluating the scale readings of an instrument in terms of the physical quantity to be measured, or of determining the corrections

to be applied to scales of any measure in order to obtain true values.

- **calibration curve**
plotting the graduation figures relating two variables progressively on a line.

- **caliper**
a measuring tool that can be set to measure the thickness of a block, diameter of a shaft, or bore of a hole.

- **calk**
to drive, force, or pour oakum, lead, mortar, or other material into joints to seal them against leakage.

- **calm**
the state or condition of the water surface when there are no wind waves or swell.

- **cam**
a device that controls or alters motion—for example, the ignition distributor breaker cam, which, in rotating, causes contact points to open and close.

- **cam chuck**
a profiling device fitted to the saddle of an ordinary lathe, to turn out irregular forms, such as cams.

- **cam engine**
engine in which the pistons are reciprocated by means of a cam and roller mechanism.

- **cam follower**
a part of an engine which rides on the contour surface of a cam and to which motion is imparted by the cam.

- **cam friction**
that friction existing between a cam and the piece it actuates.

- **cam gear**
a toothed wheel driving a camshaft in an automobile.

- **cam governor**
controlling device used in Otto-cycle gas-engines.

- **cam ground**
refers to oval-shaped piston, so ground as to permit piston to expand and assume a round shape when hot.

- **cam lobe**
the raised portions on the contour of a cam which operate the cam followers.

- **cam plate**
an eccentric rotating plate, employed to produce reciprocating motion of an intermittent or irregular character.

- **cam wheel**
  a wheel with a prcjection either on the periphery or face, adapted to give motion to another object against which it impinges, by sliding contact.
- **camber**
  a cross fall given to the surface of the top of the dam to drain off water quickly.
- **camber angle**
  the angle between the vertical and a line drawn through a front auto wheel in its up and down direction.
- **cambic horizon**
  an altered horizon resulting from movement of soil particles by frost, roots, and animals to such an extent to as destroy original rock structure.
- **cambium**
  the layer of growth cells between the wood and bark of a live tree.
- **canal**
  an artificial water channel, used for navigation purposes, for conveying water for irrigation, domestic or industrial purposes, etc.
- **canal land width**
  the width of land occupied by or acquired for a channel section including its inside berms, banks, outside borrow-pits, if any, spoil banks, if any.
- **canal scraper**
  a water-propelled scraper for removing silt and water-weeds from lined channels. The pressure of the water against a vertical plank wall, cut to fit the shape of the lining, pushes the scraping equipment down the channel.
- **canal system**
  a system comprising all the canals which constitute a complete irrigation system.
- **cane field tread**
  a type of rubber tractor tire tread found on machinery used in sugar cane fields.
- **canning**
  the process of storing food in airtight containers.
- **cannula**
  needle which is blunt.
- **canopy**
  the branches, leaves etc., formed by woody plants, some distance above the ground.

- **cant**
thick piece of timber with or without squared edges, sawn from a flitch or a log and intended for further conversion into small sizes.

- **cantilever beam**
a beam supported only at one end, which carries loads on the unsupported end.

- **cantilever bridge**
a bridge at least one portion of which projects beyond the supporting pier or piers and acts structurally as a cantilever.

- **cantilever elements**
the cantilever element, or cantilever, is that portion of a dam, which is contained within two vertical radial planes spaced a unit distance apart at the axis.

- **cantilever outlet**
an outlet spillway designed to retard water flow to prevent excessive erosion on the receiving ditch.

- **canvas**
the conveyor apron on a grain binder, combine or windrower.

- **canvas belting**
a belt made up of layers of canvas stitched together along the edges and through the centre.

- **canyon**
a deep valley with high steep slopes, generally with a stream at the bottom.

- **cap bearing**
a type of bearing consisting of two halves which are bolted together.

- **cap screw**
a screw-bolt intended to be used without a nut, used to secure the covers of small steam engines instead of studs.

- **capability class**
classification of soil that indicates the most intensive but safe use of land.

- **capacitance (C)**
a measure of the capacity to store electrical energy in an electrostatic field.

- **capacitive reactance**
resistance to the flow of AC current due to capacitance It causes the voltage in the circuit to lag the current.

- **capacitor**
a device that stores electrical energy in an electrostatic field.

- **capacitor-start capacitor-run motor**
a type of AC induction motor that develops a high starting torque while drawing a moderate starting current.

- **capacitor-start induction-run motor**
a type of AC induction motor that uses a capacitor for starting but not when running.

- **capacity curve**
the graph of the volume of reservoir, tank, etc., as function of elevations. The capacity of reservoir can only be defined by reference to some definite elevation.

- **capillary conductivity**
is denned as the grams of water that flow per second through a plane of one square centimetre area placed perpendicular to the direction of flow, where the pressure difference is one dyne per square centimetre of distance in the direction of flow.

- **capillary fringe**
the ground immediately above the water table which contains water held by capillary rise. The lower part of the fringe is fully saturated. The fringe rises higher with increasing fineness of the soil and as the water table rises. It is the lowest part of the zone of aeration.

- **capillary moisture**
water held by surface tension in the capillary spaces and as a continuous film around the particles, free to move under the influence of capillary forces, and available to plants.

- **capillary movements**
movements of soil moisture in any direction caused by capillary tension.

- **capillary porosity**
the ratio of the capillary pore space to the total volume of the soil or rock.

- **capillary water**
soil water held by adhesion forces as a continuous film around soil particles and in the capillary spaces.

- **capon knife**
a scalpel which is designed for and used in caponisation.

- **carbide insert bit**
  a detachable bit, the cutting edges of which consist of tungsten carbide embedded in a steel base.

- **carbon tool steel**
  a tool steel containing little or no-alloying metals, and with a carbon content.

- **carburettor**
  an engine fuel system component that meters and atomises the fuel and mixes it with the intake air on some spark ignition engines.

- **carburised steel**
  a low carbon steel used for plough shares to which surface carbon has been added in varying amounts and depths.

- **carolina cotton planter**
  an early type of cotton planter consisting mainly of a furrow opener, a hopper with an adjustable opening and agitating device to keep the seed flowing out of the hopper.

- **carrier drain**
  open or subsurface drain which receives the water from the relief drain and may in addition be also functioning as a relief drain for the area through which it passes.

- **cartograms**
  a type of diagrammatic and graphical representation of data wherein maps are drawn and the figures representing the phenomena at various places are shown by signs or symbols.

- **case hardening**
  a process of surface hardening involving a change in the composition of the outer layer of an iron-base alloy followed by appropriate thermal treatment.

- **Case, J I**
  early manufacturer of a machine, called the combine. The machine was used for threshing, separating, and winnowing grain crops.

- **casing head**
  in well boring, a heavy mass of iron screwed into the top of a string of casing to take the blows produced by driving the pipe home.

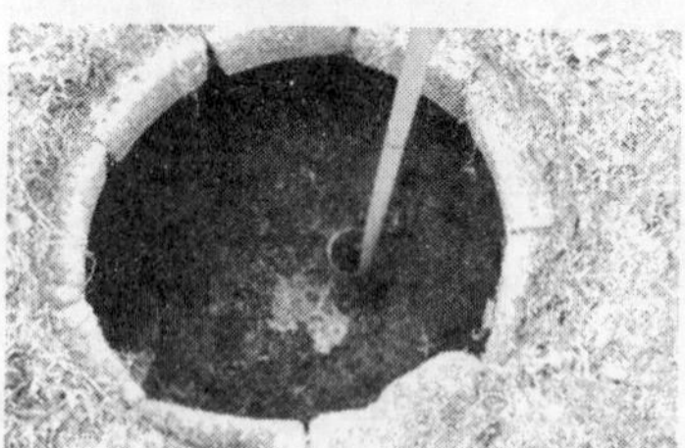

- **casing wrench**
  a special wrench for imparting to the casing the oscillat-

ing action which prevents it from freezing while it is being sunk.

- **catastrophic flood**
a flood due to meteorological conditions of exceptional rarity, and due either to catastrophic rainfall spread over a period considerably longer than the time of concentration for the area, or to violent rain of the cloud burst type.

- **catenary curve**
the curve assumed by an inelastic flexible cable of uniform cross-section and homogeneous material, when suspended between two points.

- **cavitation number**
the cavitation number is a parameter by which the experimenter can predict the onset of cavitation in the prototype.

- **cavity tubewell**
a tubewell, which being without strainers, draws its supplies from a cavity developed in the water-bearing aquifer at the end of the pope.

- **ceiling leaf area index**
leaf area index at which mutual shading of leaves in a canopy is complete and net photosynthesis is zero, i.e., gross photosynthesis equals the respiration losses of the whole crop.

- **celery harvester**
a hand pushed or tractor-pushed device with two or more 'L' shaped teeth which turn inward on either side of the row and which cut the roots of the celery preparatory to harvesting.

- **cellular coffer dam**
a water coffer dam built with steel pile sheeting driven either in the form of semi-circular cells or circular cells, which are joined by connecting arcs of steel piles. The cells are either filled with sand or earth and are braced; usually supported by berms.

- **cemented soil**
a soil in which the grains or aggregates adhere firmly and are bound together by some material that acts as a cementing agent such as colloidal clay, iron, silica, or alumina hydrates, lime, etc.

- **central limit theorem**
regardless of the shape of the original population, the sampling distributions of commonly computed statistics (e.g., distribution of means, standard deviations) tend to be approximately normal.

- **central mounted tractor mower**
a type of tractor mower in which the cutter bar is mounted between the front and rear wheels of the tractor and to the right of the operator.

- **central point distribution system**
a common electrical distribution system in which electrical service for dispersed load centres originates from one central location.

- **centralised farrowing**
a handling system in which sows are placed in individual pens, in a permanent building, during the time they are farrowing, as opposed to portable house farrowing.

- **centre pivot irrigation**
the world's first successful automatic irrigation machine, likely to increase the demand for water considerably in warm dry regions.

- **centre punch**
steel punch with one end ground to a point; for lying out work.

- **centre shovel**
the shovel mounted between the front and rear shovel of a cultivator gang. Such gangs usually have only three shovels: front, centre and rear.

- **centrifugal pump**
a rotodynamic pump in which a bladed impeller drives the water outwards so that it leaves the casing radially.

- **centrifugal switch**
a switch opening by centrifugal force.

- **centripetal pump**
a pump with a rotating mechanism that gathers a fluid at or near the circumference of radial tubes and discharges it at the axis.

- **cetane number**
a measure of the ignition quality of diesel fuel.

- **CF horizon**
a C-horizon which is disintegrated but not chemically transformed.

- **chain belt**
a belt consisting of links of metal used for power transmission. It is very solid but flexible chain.

- **chain gauge**
device consisting of a tagged or indexed chain or other line attached to a weight which is lowered to the water surface, whereupon the gauge height is read on a graduated staff or

opposite an index. Especially suited to bridges.

- **chain pump**
a device for raising water through small lifts consisting of an endless chain dipping below the water to be raised and running over a drum of wheel by which it is moved. In the most common form the chain has at intervals discs or lifts which fit the tube through which the ascending part passes and which carries the water.

- **chain scraper**
a chain, rather than blades, that scrapes the sludge along the bottom of a radial-flow sedimentation tank.

- **chance causes**
in the quality of a product process or service, where there is a stable pattern of variations that can be considered natural and inherent in the process, the variations are said to be due to chance or random causes.

- **channel history**
a statement or sheet showing in chronological order, the behaviour of a channel regarding its performance in the past-silt clearance done and its frequency, changes in masonry works effected and their reasons, changes in hydraulic elements of the canal (bed, width, depth, slope, discharge) and their reasons.

- **channel roughness**
the roughness of a channel, including the extra roughness due to local expansion or contraction and obstacles, as well as the roughness of the stream bed proper, i.e., friction offered to the flow by the surface of the bed of the channel in contact with the water. It is expressed by the roughness coefficient of the velocity formulas.

- **channel terrace**
type of contour terrace for soil conservation; wide and relatively shallow channel excavated at suitable intervals on a falling contour with a suitable longitudinal gradient; excavated soil deposited as a wide, low ridge along the lower ridge of the channel.

- **chanoine wicket dam**
a movable dam, consisting of a water curtain formed by timber leaves or wickets inclined downstream; the butt of the wicket is supported against a sill along the upstream side of the base slab and

the wicket is pivoted a little below its centre point to a collapsible horse held in place by a long prop.

- **charpit method**
a method of removal of stumps by burning.

- **cheated wheel tractor**
a compactor for use in controlled tipping.

- **check**
a farm structure, permanent or temporary, built, put or inserted in the ditch to check partly or wholly the water flow, so as to regulate and raise the water level in the field ditch in order to permit diversion from the ditch into one or more smaller field ditches or to deliver water on the land through cuts in the banks or through levee gates.

- **check dam**
small low dam constructed across a gully or other watercourse in order to reduce velocity of water flow for minimising channel scour and inducing deposition of eroded material.

- **check drop**
when drop is combined with a check, the dual-purpose structure is called check drop.

- **check fork**
a device on a check row planter which strikes the button on the check wire to open the valve in the planter so as to drop the seed.

- **check irrigation**
application of a comparatively large stream of water in level plots surrounded by levees.

- **cheesy**
having a more or less elastic character, deforming considerably without rupture, yet broken without difficult or the application of much force (Characteristics of certain highly colloidal soils when thoroughly wet).

- **chemical erosion**
removal of material following dissolution or chemical reactions which have put it in soluble or gaseous form.

- **chemical grouting**
a process in which chemicals are injected under pressure into a pervious medium to make it watertight by formation of gels.

- **chemical potential of water**
the chemical potential of a substance in a system is a measure of the capacity of that sub-

stance to work. In a simple solution of a non-electrolyte in water, the chemical potential of water depends on the mean free energy per molecule and the concentration of water molecule.

- **chemurgy**
  chemical research designed to promote the use of agricultural products for industrial purposes, or to obtain new, valuable products from plants.
- **chilled casting**
  casting made in an iron or steel faced mold.
- **chip test**
  a method of identifying metals based on the nature of a chip produced by a cold chisel.
- **chisel plough**
  a type of plough consisting of several beams which stir the soil without turning up the moist sub-soil or turning under any crop residue which may be found on the surface.
- **chisquared**
  the ratio of an observed sum of squares to the appropriate or corresponding variance as fixed by a hypothesis.
- **chisquared statistics**
  a statistic which is distributed as chi-squared (X^2), namely as the sum of squares of independent standard normal variant. It is widely used to test agreement between observation and hypothesis.
- **chisquared test**
  the test of significance based upon the chisquared statistics.
- **chocolate mousse**
  water-in-oil emulsion naturally emulsified at sea from viscous oil.
- **chopper**
  device consisting of a rotating commutator to interrupt continuous wave audio frequency signals at either the transmitter or receiver.
- **chromatography**
  a method for the separation of compounds from one another.

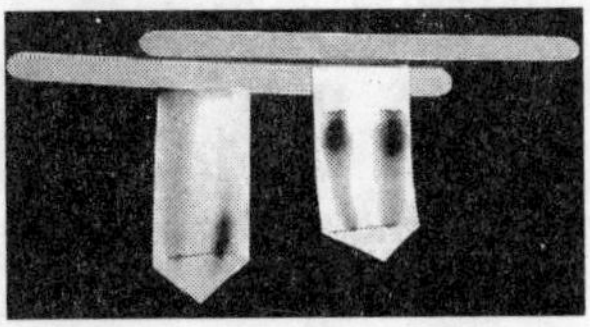

- **churn barrel**
  a vessel in which cream is agitated until butter is formed.
- **churn drill**
  a drill for drilling wells by the use of cable tools.
- **cinch band or belt**
  plies of cord material under the treat area of a tire having the

cores nearly parallel to the centre line of the tire. These cords do not tie into the tire beads but furnish circumferential strength for the tire.

- **circuit**
  a complete path for the flow of electrons.

- **circuit breaker**
  a device commonly used to protect electrical circuits and loads from over current conditions.

- **circuit overload**
  surplus current flow in an electrical circuit.

- **cladophora, blanket weed**
  stringy green algae commonly found in nutrient rich waters. They mat into a green blanket in the water.

- **claiming pen**
  a pen or small enclosure within a pen in which an ewe and her newborn lamb can be placed until the ewe accepts the lamb as her own.

- **clamp-on ammeter**
  a type of ammeter that can be used without connecting it directly into the electrical circuit.

- **Claus process**
  extraction of the element sulphur from the gas hydrogen sulphide (H_2S) by passing it mixed with air at about 150°C over an iron or bauxite catalyst. The exhaust gas (tail gas) consisted of SO_2, H_2S, COS (carbonyl sulphide) and CS_2 (carbon disulphide) corresponding to some 15000 p.p.m. of SO_2, but it has recently been reduced to about 250 p.p.m. with an increase in the recovery of sulphur.

- **clay humus complex**
  the close association between the colloids of clay (mineral) and finely divided particles of humus (organic) in the soil. The clay humus complex is chemically active in capturing exchangeable nutrients, which are then available for plants.

- **clay loam**
  a clay loam is a fine textured soil which breaks into clods or lumps which are hard when dry. When the moist soil is pinched between the thumb and finger it will form a thin ribbon, which will break readily, barely sustaining its own weight. The moist soil is plastic and will form a cast, which will bear much handling. When kneaded in the hand it does not crumble readily but tends to work into a heavy, compact mass.

- **clear cutting**
cutting all trees down to prepare land for cultivation.
- **clear span structure**
a building with no centre posts.
- **cleat wiring**
a method of supporting wiring on porcelain insulators or cleats. By the use of cleats both wires of a circuit are held at the correct distance apart.
- **cleated conveyor**
a rubber belt conveyor to which upstanding cross-strips (cleats) of metal or rubber are riveted or welded, enabling it to carry material up steep slopes.
- **closed circuit**
a complete pathway from the voltage source to the load(s) and back to the voltage source.
- **colluvial**
deposited soils due to gravity.
- **combine**
a mechanism used for threshing and harvesting grain.
- **combustion**
the chemical process of oxidation that produces heat.
- **commodities**
agricultural products such as corn, wheat, livestock, etc.
- **compass**
a mechanical drafting device used for drawing circles, arcs and ellipses.
- **compost**
mixture of partially decayed organic matter.
- **compound pressure relief valve**
a pressure limiting hydraulic valve that operates in two stages.
- **compression**
1. a reduction in the volume occupied by a substance.
2. the second stroke of the four-stroke engine cycle.
- **compression gage**
a tool used for measuring the pressure developed in an engine combustion chamber.
- **compression ignition engine**
1. a kind of engine in which the fuel is ignited by the heat of compression in the combustion chamber.
2. another name for a diesel engine.
- **compression ratio**
the ratio of the volume of the combustion chamber with the piston at BDC to the volume with the piston at TDC.

■ **compression strength**
the ability of a material to withstand a compressive force.

■ **compression, rings**
piston rings that are designed to seal between the piston and cylinder wall to prevent loss of pressure in the combustion chamber.

■ **compressor system**
the primary internal combustion engine system that is involved in creating compression and producing power from the combustion of an air-fuel mixture.

■ **computer assisted design (CAD)**
the use of computer hardware and software to produce drawings.

■ **concrete**
a rock-like construction material which is formed from a mixture of paste and aggregates.

■ **concrete masonry building**
a construction built chiefly from concrete blocks.

■ **condenser**
an engine ignition component that functions as a capacitor.

■ **conduction**
the transfer of heat through the interaction of molecules or atoms without the conducting material itself being moved.

■ **conductor**
a substance that has low resistance to the flow of electrical current.

■ **conduit**
a channel or tube through which electrical conductors are run in order to provide the conductors with mechanical protection.

■ **conduit regulator**
a tank with a free water surface provided at the transition from tunnel or pipeline to penstock to protect the tunnel or pipeline from excessive changes in pressure, rise in pressure in particular and which expeditiously maintains equilibrium in the system by strong or furnishing water required for the above purpose.

■ **cone crusher**
a crusher, used as secondary or tertiary crusher, capable of producing large quantities of uniformly finely crushed stone.

■ **cone grate**
the protective cover of an inlet riser which is slatted to pre-

vent the entrance of foreign matter.

- **cone of depression**
a cone shaped depression in either the water table in unconfined flow or the piezometric surface in confined flow developed around a well which is being pumped at any given rate of discharge.

- **cone of influence**
the depression, roughly conical in shape produced in a water table, or other piezometric surface, by the extraction of water from a well at a given rate.

- **confetti**
disks or strips of paper, ribbon, or tape, to indicate surface flow patterns in models.

- **confidence belt**
the area between the upper and lower confidence limits is the confidence belt.

- **confidence level**
a statement of the confidence with which a statistically based statement may be made, equal to 1.00 minus the confidence level. Thus if the probability that a parameter is beyond certain limits is 0.05, the CONFIDENCE COEFFICIENT that the parameter is within the limits is 1.00—0.05 or 0.95.

- **confidence limits**
the statistics t_1 and t_2 which define the upper and lower limits of the confidence interval are called the confidence limits or the fiducial limits. The confidence limits are statistics computed from samples and, therefore, subject to sampling fluctuations, will vary from sample to sample, but the parameter to be estimated by them remains fixed.

- **confined aquifer**
an aquifer found between impermeable layers above and below it.

- **confined eddy**
an eddy confined in contact with a work, such as a guide bank.

- **confined flood**
a term applied to a hypothetical flood in case of a breach or crevasse in a certain reach of a river that would have reached the lowest point had there been on breach.

- **confined flow (system)**
flow of confined ground water.

- **confined groundwater**
groundwater that is overlain not by unsaturated ground, like unconfined groundwater,

but by impermeable soil or rock.

- **confounded and fractional replication**
a factorial experiment in which only some of the treatment combinations occur in each block. Particularly used when several factors each at 2 or 3 levels are present.

- **congellipedoturbation**
it is the mixing of soil by freeze thaw cycles as in patterned ground of tundra and alpine landscapes.

- **conglomerate**
rock consisting of round water borne pebbles cemented in a matrix of fine material, calcium carbonate, silica, iron oxide etc., consolidated gravel or shingle.

- **conjunctive irrigation planning**
planning an irrigation project in a given area having underground resources of water supply in addition to the over ground, so that both the surface and subsurface water resources are so planned as to obtain the most economical and wise combination of the use of water from the two sources.

- **conjunctive use of resources**
the integrated use of resources of water.

- **connate water**
water in deep sedimentary strata that has been there since they were deposited.

- **connected load**
the total amperage of all loads connected to a building's service entrance panel, usually calculated on a 240-volt basis.

- **connecting rod bearings**
the bearing at end of an engine connecting rod which attaches to the crank shaft throw.

- **connector**
a fitting used to join one piece of hydraulic piping to an-

other piece of piping or to a hydraulic component.

- **conoidal dam**
  a dam which can be considered as being formed by the juxtaposition of a certain number of blocks.

- **consequent lake**
  a lake that exists in consequence of the original inequalities of a land surface.

- **consequent stream**
  a stream whose course has been controlled in direction and location by the general slope of the surface topography.

- **conservation**
  the act of keeping entire, of protecting, preserving and improving with undertones of permanence and saving from waste (in quantity); and deterioration (in quality); also restoration to original state and improvement of what has been wasted or deteriorated.

- **conservation cropping**
  a way of farming that aims to maximise the protection against erosion that can be achieved through soil and crop management for sustained farm productivity.

- **conservation farming**
  the most efficient use of the land over a long period of time which protects and preserves the land for agriculture.

- **conservation pool**
  the storage space in a reservoir which cannot be drained below the outlet of a dam.

- **conservation reservoir**
  a reservoir, impounding water for useful purposes, such as municipal supply, power, irrigation and recreation.

- **conservation storage**
  storage impounded for later release, as required for some useful purposes, such as municipal supply, power or irrigation.

- **conservation tillage**
  method of soil preparation, planting and cultivation that disturbs the soil the least and leaves the maximum amount of plant residue on the surface.

- **conservatory**
  a glass structure constructed for the accommodation of flowering plants for display as contrasted to a structure for propagation which may or may not be attached to a house.

- **consistency meter**
  meter to indicate changes in the consistency of the mix and to provide information of maintaining uniformity in the concrete.

- **consistency, soil consistency**
  a term expressing the degree of cohesion of the soil particles and the resistance offered to forces tending to deform or rupture the aggregates.

- **consolidated drained test**
  a soil test in which essentially complete consolidation under the confining pressure is followed by additional axial (or shearing) stress applied in such a manner that even a full saturated soil of flow permeability can adapt itself completely (full consolidated) to the changes in stress due to the additional axial (or shearing) stress.

- **consolidated undrained test**
  a test in which complete consolidation under the vertical load (in a direct shear test) or under the confining pressure (in a trail test) is followed by a shear at constant water content

- **consolidated water rate**
  a method for charging for the supply of irrigation from Government canals.

- **consolidation test**
  a test in which the specimen is laterally confined in a ring and is compressed between porous plates.

- **consolidation time curve**
  a curve that shows the relation between (1) the degree of consolidation and, (2) the elapsed time after the application of a given increment of load.

- **constant angle arch dam**
  it is a type of an arch dam in which all the arch rings have a constant value of central angle.

- **constant error**
  an error that consistently affects all observations or measurements by an equal amount in the same direction and that, accordingly, introduces a systematic bias into the observations or measurements.

- **constant head orifice offtake regulator**
  a flow regulating structure of small channels built where

sufficient head is available between the parent channel and the off take, consisting essentially of a stilling pool and two gates, the adjustable orifice gate and the turnout gate placed on the upstream and downstream of the pool respectively.

- **constant radius arch dam**
  an arch dam the arch centres of which for the upstream face, the downstream face and the centre line are all coincident with the axis centre at all elevations.
- **constant running power take off**
  a type of power take off having its own clutch which allows it to continue to operate when forward motion of the tractor is stopped.
- **constant spring**
  a spring whose variation in discharge from maximum to minimum does not exceed about one-third of the average discharge.
- **constant upstream level offtake regulator**
  a structure designed to deliver, automatically, a constant discharge when operated in conjunction with auxiliary equipment, or a structure to control the upstream water level that is not affected, within limits by changes in downstream level.
- **constant upstream level outlet**
  outlet or turnout in which a gate installed in the supply channel keeps the water constant or nearly constant upstream of the outlet.
- **constriction**
  a short length of a closed conduit with a flow section less than the cross-section of the main conduit, in the form of a partition or diaphragm or a throat, with the cross section varying along the conduit axis.
- **construction drawings**
  drawings which give in detail the whole plan to the proposed construction.
- **construction joints**
  they are breaks in the continuous pouring of concrete required to prevent haphazard cracking of the structure. These are classed into (1) horizontal joints, (2) transverse joints and, (3) longitudinal joints.

- **consumptive irrigation requirement**
is the depth of irrigation water exclusive of precipitation, stored soil moisture or ground water, that is required to meet evapotranspiration demand of the crop.

- **consumptive use efficiency**
the ratio of consumptive water use by the crop of an irrigated farm or project and the irrigation water stored in the root zone of the soil on the farm or the project area.

- **contact aerator**
a variety of submerged bed aeration with a swinging diffuser for air and no solid packing.

- **contact herbicide**
an herbicide that kills primarily by contact with plant tissue, as distinguished from translocated herbicides.

- **contact placement**
the drilling of seed and fertiliser together while sowing.

- **contact spray**
a solution containing a herbicide or insecticide that kills on contact.

- **container grown**
plants that are grown in pots or other type of container and are shipped in the container.

- **contaminate**
addition of such substances that change the purity or usefulness of a substance.

- **contemporaneous erosion**
erosion of a local or unimportant character that goes on while sedimentation is taking place quite generally elsewhere.

- **continental basin**
a region in the interior of a continent comprising one or several closed basins.

- **contingency cropping**
contingency cropping is growing of crops in aberrant situations like drought and floods.

- **continuous action atomizer**
a type of hand spray consisting of a pump and a chamber

in which the spray liquid is poured. The pump builds up enough pressure after a few strokes to force out a continuous stream of spray.

- **continuous apron**
any type of continuous belt used to convey material from one point to another.

- **continuous bucket elevator**
an elevator with a continuous series of buckets attached to a belt or chain.

- **continuous cropping**
cultivation of the same crop on the same land year after year.

- **continuous delivery**
in irrigation a system by which each irrigator receives his allotted quantity of water at a continuous rate.

- **continuous demand**
demand incident on a year round basis, as contrasted with seasonal or other periodic demand.

- **continuous drying**
drying carried out with a continuous flow of grain and air in contrast to batch operation; used in connection with conditioning of harvested produce.

- **continuous interstice**
an interstice occurring in granular material, or a fracture in rock formation, or a tubular opening in soluble rocks or lava which is connected with other interstice.

- **continuous irrigation**
applying water to the soil for a long period of time, as used in frost protection, water is applied continuously as long as the frost hazard exists.

- **continuous ploughing**
method of ploughing, first the headland is marked and the first ridge is set up at three quarter of land width from the side. The outer ridges are set at full width over the field. The operator starts ploughing between the first ridge and the side land, continue to turn left and cast in the three quarter land until a quarter land width of ploughing is completed on each side.

- **continuous signal**
an electrical signal that can vary between a range of values.

- **continuous stave pipe**
a pipe line made of wooden staves so constructed that the staves forming the barrel of the pipe are placed in a con-

tinuous line from one end of the pipe line to the other; it is assembled in place in the field with staggered joints and held to position and tightness by encircling steel bands.

- **contour basins**
basins which are made by levees or borders following contours, with occasional cross levees.

- **contour border irrigation**
the application of water to a very gently sloping field which has been divided into strips by earth ridges on the contours.

- **contour bunding**
the construction of small bunds across slope of the land on a contour so that the long slope is cut into a series of small ones and each contour bund acts as a barrier to the flow of water.

- **contour cropping**
the cultivation of crops along the contours of a slope.

- **contour fence**
a fence following an approximate contour or elevation, but may be composed of a number of curved and straight sections.

- **contour line**
a line drawn on a map showing heights above sea level.

- **contour map**
a map showing the configuration of the surface by means of contour lines drawn at regular intervals of elevation as one for every twenty feet, a crowding of the contour lines indicating steepness.

- **contour practice**
operations such as ploughing, dishing, planting, cultivating and harvesting across the slope and on the level.

- **contour strip cropping**
the farming of sloping land in alternate strips of inter-tilled row crops and close growing crops.

- **contour trenching**
construction of trenches along a uniform level at a particular contour across steep slope areas.

- **contract bed**
an artificial bed of coarse material, such as broken stone or clinkers, in a watertight basin provided with controlled inlet and outlet.

▪ **contracted measuring weir**
a weir whose crest extends only part way across the channel in which it is installed and is terminated by extensions in the same plane as the crest, these extensions rising above the water level on the upstream side of the weir, their effect being to produce a contraction in the width of the stream of water as it leaves the notch.

▪ **contraction joint grouting**
the grouting process whereby contact between a structure and a vertical or nearly vertical rock face is grouted.

▪ **control band**
the band between the control limits, as specified may include 99.7 per cent of the observations or samples.

▪ **control flume**
an open conduit or artificial channel arranged for measuring the flow of water, generally including a constricted section wherein critical depth exists.

▪ **control structure**
a stage discharge regulating device of the spillway. It may be of any form, viz., weir, side channel, glory spillway orifice, tube, pipe or a channel.

▪ **controlled atmosphere**
adjusting carbon dioxide and oxygen where food is stored and while it is being transported.

▪ **controlled concrete**
concrete made of selected materials, accurately measured, thoroughly mixed, and having a constant water cement ratio and consistency, all batches being as nearly uniform as possible.

▪ **controlled mosaic**
a mosaic in which the photographs have been adjusted by reference to ground control so as to produce a continuous photographic representation of a part of the earth's surface.

▪ **convection**
the transfer of heat to or from an object by a gas or liquid.

▪ **convection dryer**
a dryer that works by the convection of hot gas, usually air.

▪ **convection oven**
an oven which heats food by circulating hot air.

▪ **conventional theory**
the theory stating that electricity is the flow of posi-

tively charged particles through a conductor, and that flow is from positive to negative.

- **conventional tillage**
  land is ploughed, turning over all crop residues.
- **converging tube**
  a tube whose diameter decreases, usually at a uniform rate, along its longitudinal axis in the direction of flow.
- **conversion factor**
  that factor by which the numerical value made in one system of units must be multiplied to arrive at the numerical value in another system of units.
- **conveyance culvert**
  a culvert which carries canal flow under road, railroads or other canals.
- **conveyor**
  a machine equipped with belts, cables, chains, buckets, etc., for the transportation of materials.
- **cool house**
  a house of range of greenhouses which is maintained at a day temperature of about 55° to 60° and a night temperature of about 50° to 55°.
- **coolant hydrometer**
  a tool for measuring the freeze protection of an engine coolant mixture based on the specific gravity of the mixture.
- **cooling tower**
  a tower at a power station or other industry with large boilers or generating capacity, used for the air-cooling of hot water by evaporation.
- **corn picker**
  apparatus that removes ears of corn from stalks.
- **corrosion resistance**
  the ability of a material to resist oxidation.
- **cosine**
  a trigonometric term used for indicating the ratio of the adjacent side of a given acute angle in a right triangle to the hypotenuse.
- **cotton gin**
  an apparatus that removes cotton seed from cotton fibre.
- **counter voltage**
  an induced voltage opposing any change in the current flow in a circuit.
- **covalent bonding**
  the type of chemical bond between two atoms when they share electrons in pairs so that

each atom provides half of the electrons.

- **cracking pressure**
the pressure at which a hydraulic pressure relief valve first begins to open and divert oil flow to the reservoir.

- **cradle scythe**
a hand-held grain harvesting tool having a long, curved blade attached to a long, bent handle.

- **cranking amperage**
electrical current flow required for starting an internal combustion engine.

- **crankshaft throw**
offset on an engine crankshaft that converts linear piston movement to rotary crankshaft motion.

- **crumber**
a bulldozer blade that follows the wheel or ladder of a ditching machine to clean and shape the bottom of the drain trench.

- **crushing and pressing**
the process whereby juices or saps are separated by pressure from the less valuable fibre or pulp.

- **crust mulch structure**
an arrangement where a surface crust is underlain by a horizon of loose, inco-herent particles of mealy, crumb, or granular structure.

- **cryoscope**
an instrument which is used to determine the freezing point of milk.

- **crystal lattice**
arrangement of atoms in an orderly, repeating, three-dimensional pattern.

- **cubing**
process of forming hay into high density cubes to facilitate transportation, storage and feeding.

- **cubing machine**
a device which macerates meat to make it tender.

- **culti cutter**
an implement with heavy spade like blades which is used for cultivating the sod cover maintained in orchards.

- **cultipacker**
a machine used to brush clods and firm or pack the soil utilising one or more rollers made up of V-shaped wheel sections or those having sprocket-like teeth on their periphery, so as to present a corrugated or jagged working surface to the soil.

cultivation
includes any labour and care taken in the raising of plants such as stirring soil, fertilizing, etc., or to loosen the soil around a plant for the primary purpose of weed control.

cultivator teeth
a general classification for all types of shovels and sweeps used on cultivating equipment.

culvert
a transverse drain or waterway structure under a road, railroad, canal, or through an embankment by means of a pipe or enclosed channel.

cumulative distribution curve
the plot of the cumulative frequencies for a particular class interval and of all lower class intervals below it (or of all upper class intervals above it), against the class interval concerned.

cumulative frequency curve
a frequency curve in which values are added successively to each other and then converted into percentage.

cup basin type water seal
a water seal made effective by dipping the lower limb of the hood into a cup-like basin provided on the downstream end.

cup drill
a grain drill with vertical seed-metering wheels with cups on the edges which lift up through a seed pool and throw approximately one seed per cup over into the seed tube.

cupola
a furnace in which iron is melted to make iron castings. Cupolas emit grit, ash and sometimes oil vapour.

curd mill
a machine having a multiple series of knives which is used for cutting curd in Cheddar cheese making.

curtain grouting
it consists of forming an approximately vertical grout curtain of moderate thickness below adjoining a hydraulic structure to reduce the passage of water to a required degree.

curve of demand
a graph showing the amount of water needed for irrigation

at various times during the crop season based on elements of time and quantity.

- **curve of supply**
a graph showing the water available based on elements and time and quantity.

- **curved wing**
a device to prevent excess bottom sediment entering an offtake channel and to ensure more or less proportional distribution of sediment between the parent and offtake channels.

- **cut water**
the upstream nose of a pier shaped for streamline entry of flow into bays.

- **cutoff drainage system**
a tile drainage system normally placed near the upper edge of a wet area to intercept water before it gets down the slope.

- **cutover**
an area from which some or all the trees have been removed.

- **cutstone**
a stone that is cut or finished accurately to definite shape and size.

- **cutter**
a machine or person who cuts or chops.

- **cutter dip planter**
an equipment for planting cane sets fitted with a spraying attachment to allow chemical treatment of sets. Fertiliser distributors and compaction rollers are also fitted if desired.

- **cutterhead**
the device on a silage harvester which chops the forage into suitable length for silage.

- **cutting**
vegetative part removed from the parent plant and managed, so that it regenerates itself.

- **cutting fluid**
a liquid used in many metal cutting operations to keep the saw blade and metal cool, clean and lubricated.

- **cycle**
a complete set of events recurring in the same sequence.

- **cycle efficiency**
the ratio of the heat out to the heat in.

- **cycle of erosion**
course of development in evolution of landscape comprising of stages youth, maturity and old age.

- **cycle of fluctuation**
in ground water hydrology the total time occupied by a

period of rise and succeeding period of decline of a water table.

- **cycle time**
  the amount of time required for one complete set of events to occur in a hydraulic system.

- **cylinder block**
  primary engine structural component housing the reciprocating engine components.

- **cylinder leakage test**
  a test used to determine the extent and source of engine compression losses.

- **cylinder teeth**
  the teeth attached to a spike-tooth type cylinder in a grain thresher or combine.

- **D horizon**
  unweathered rock below the c horizon.

- **dam**
  a barrier across a watercourse to impound water or create a reservoir (a) for raising the water level, (b) for diverting water into a conduit or channel, (c) for creating a hydraulic head that is then used to generate power, (d) for improving river navigability brought about by controlled releases of stored water, and (e) for retaining debris.

- **damping off**
  young seedlings that are at the ground level are infected by this disease, which causes them to rot and fall over.

- **datum**
  an arbitrary horizontal line which is used in the calculation of heights and depths, as used in a longitudinal section of a canal, rail-road plan.

- **DC motor**
  a motor designed for operating on direct current electricity.

- **dead finish**
  the line in a field which acts as the meeting point of two directions of ploughing, wherein the slices are turned away from each other.

- **dead furrow**
  a double furrow that is left between two lands or in the middle of the field while ploughing.

- **dead load**
  the weight of the structural components of a building.

- **dead storage**
  the amount of water at the bottom of a reservoir, which cannot be drawn out due to it

being lower than the outlet or because it is dirty, or for other reasons.

- **debris avalanche**
the sudden down slope movement of the soil mantle on steep slopes, brought about due to the complete saturation of the soil through protracted heavy rains.

- **debris dam**
a barrier constructed across a stream or channel for stopping transport of debris along the bed.

- **decking**
boards or planks used to form the facing or shutters to concrete floor slabs.

- **deckless buttress dam**
a type of buttress dam in which the face slab is replaced by flaring the upstream edges of the buttresses to span the distance between buttress walls. The flaring portion of the buttress (head) may be of various geometrical shapes, viz. (1) massive head, (2) round head or mushroom head, (3) diamond head.

- **decompaction**
decrease in soil density (dedensification), brought about by over-vibration, for example.

- **deep bin**
a bin that is deep in relation to its width; if filled with grain the lateral pressures on opposite sides of the bin react with each other and affect the force relations in the design of the bin.

- **deep placement**
placement of fertiliser at some depth, so that the full depth of the root zone is enriched with nutrients.

- **deep ploughing**
an operation intended to mix soil below the depth of normal tillage with the normally tilled layer or to cover undesirable soil surface layers of sub-surface soil.

- **deep well pump**
a pump used for lifting water from deep wells, the pumping mechanism usually being installed within the well at a considerable depth below the surface.

- **Deere, John**
blacksmith and manufacturer of the self-scouring steel plough.

- **deflation**
  the complete removal of loose material by the wind, leaving the rocks bare to the continuous attack of the weather.

- **deflocculate**
  to separate or break down soil aggregates into their component particles.

- **degenerated soil**
  a soil which, after having reached maturity in a certain line of development, is subjected to new conditions of plant cover, drainage or climate, and loses all or some of its character.

- **degraded alkali soil**
  a soil containing less than 0.1 per cent salt but with appreciable quantities of exchangeable sodium and hydrogen. Soil pH values in the surface horizons may be below 6.0 but are usually between 6.5 and 8.0. This corresponds approximately to the great soil group known as SOLOTH.

- **degree of discharge control**
  the total usable release in per cent of the total demand.

- **degree of dispersion**
  extent to which soil aggregates are broken down to primary particles by any given treatment.

- **degree of drainage**
  a qualitative measure of the efficacy of a drainage system in an area.

- **degree of duration control**
  the per cent of the total period of study during which releases are at a rate equal to or exceeding the rate of demand.

- **dehulling**
  the removal of husk or hull (outer covering) from intact grains or seeds.

- **dehydrator**
  a device used for drying food.

- **de-hydro-frozen**
  the process of removing moisture after partial cooking and then freezing.

- **demand load**
  the maximum amperage electrical load expected to operate at any one time, usually calculated on a 240-volt basis.

- **depletion region**
  a region depleted of electrical carriers.

- **design function**
  the principle for which a structure has been created.

- **dielectric**
  the insulating stuff separating the plates of a capacitor.
- **differential cylinder**
  a hydraulic cylinder having unequal piston areas exposed to fluid pressure during the extension and retraction strokes.
- **digital integrated circuit**
  an electronic circuit that produces, amplifies, or responds to signals having two values - on or off.
- **direct acting pressure relief valve**
  a pressure limiting hydraulic valve in which fluid pressure acts directly against a spring loaded check ball.
- **direct current (DC)**
  a type of electricity having a constant voltage and polarity with respect to time.
- **disconnecting means**
  a device for opening the circuit between all ungrounded electrical conductors and a motor and motor controller.
- **discrete semiconductor component**
  an individual electronic component that is made from semiconductor material.
- **divider**
  mechanical drafting tool used to transfer measurements and spacing from one location to another.
- **dome**
  top of a piston. It is also called the piston head.
- **doping**
  adding elements to pure silicon to alter its electrical characteristics.
- **double-pole circuit breaker**
  a type of circuit breaker that simultaneously opens or closes two ungrounded circuit conductors.
- **draft**
  animals used for work.
- **drafting machine**
  a mechanical drafting device that takes the place of the T-square and triangles.
- **drawbar power**
  tractor power available for pulling a load attached to the drawbar.
- **drawing board**
  a wooden board to which drafting paper is attached in order to make mechanical drawings.
- **ductility**
  the degree to which a metal can be cold.

- **duplex convenience outlet (DCO)**
an electrical device having two receptacles on a single strap.

- **duration**
the length of time that an engine valve is open. It is based on the number of degrees of crankshaft rotation.

- **duty cycle**
the relative amount of a 10-minute period during which a welding machine can operate at its rated amperage.

- **dwell**
number of degrees of crankshaft rotation that mechanical breaker points are closed.

- **eagle claw**
a cultivator gang with four shovels which are arranged in the shape of an eagle's claw.

- **earth (dike) coffer dam**
a type of water coffer dam built entirely with earth materials protected by a revetment of sand bags.

- **earth dam**
a barrier of earth, clay, sand or sand and .gravel, or a combination of the earth and rock.

- **earth lining**
a thin covering of highly impervious soil placed in canals or other retention reservoirs to reduce seepage losses.

- **earth loader**
a machine consisting of a frame mounted on crawler tracks, a cutting blade, mouldboard plough, conveyor chute, conveyor belt, and an engine to operate the belts; used to load large hauling units.

- **earth moving**
tillage action and transport operations utilised to loosen, load, carry and unload soil.

- **earthfast**
a well is said to be earthfast when the friction of the earth against the outside of the drum becomes so great as to stop its descent before the requisite depth is attained.

- **earthing up**
the process of putting the earth or soil just near the base of stems of certain crops like sugarcane, cassava, papaya, potato, banana etc., to give support to the plants.

- **ebbing and flowing spring**
a spring which has periods of relatively large continuous discharge occurring at more or less regular and frequent

intervals. Such springs may be classified as perennial or intermittent.

- **echard**
the water present in the soil, which is unavailable to plants.

- **eco-farming**
the potential for introducing mutually reinforcing ecological approaches to food production.

- **ecological factor**
anything in the environment which affects the growth development and distribution of plants, and hence determines the characteristics of the plant community.

- **economic life of a project**
the life of a project determined by the point in time at which the effect of economic and physical factors (physical depreciation, obsolescence, changing requirements for project services, and time discounts and allowances for risk and uncertainty) is to cause the costs of continuing the project to exceed the additional benefits to be expected from continuation. The economic life of a project is generally less than its physical life. The period of analysis can be shorter or equal to economic life of a project.

- **economic yield**
the maximum rate at which water can be artificially withdrawn from an aquifer throughout the foreseeable future without depleting the supply or altering the chemical character of the water to such an extent that withdrawal at this rate is no longer economically possible.

- **economical rate of runoff**
the critical rate of runoff for which it is economically feasible to provide a control channel in drainage schemes.

- **ecosystem**
an ecological system, one in which there is a cycle of interchange of materials between living organisms and their environs. An ecosystem includes the plant and animal communities in an area together with the non-living physical environment that supports them. The boundaries defined by ecosystems are dynamic, as they can change with time.

- **ecosystem development strategy**
  a planning strategy (e.g. a lake) that would reduce the nutrient content of sewage and any other effluent flowing into it.

- **ectodynamorphic soil**
  soil shaped by influences other than parent material.

- **edaphic**
  a term referring for the soil conditions or types as ecologic factors. Pertaining to or influenced by soil conditions.

- **edaphology**
  the scientific study of the relationship between soils and living entities, including man's use of the land.

- **edaphon**
  the whole living community of the soil.

- **edge drop seed plate**
  one of the types of seed plates used for planting corn which carries the corn kernel on its edge in the cell of the plate.

- **edge of well**
  outer periphery of well tubes including screen.

- **eductor**
  a device fitted to a hose near its mouth, enabling a chemical to be sucked in, in the correct proportions.

- **effective field capacity (tillage)**
  the actual area covered by the implement based on its total time consumed and its width.

- **effective irrigation**
  controlled and uniform application of water to crop land in required amount at required time, with minimum cost to produce optimum yields without waste of water and adverse effect on soil in form of soil salinity and water-logging problems.

- **effective length of slope**
  that length of slope where there is no interception and water may build up to erosive flow.

- **effective porosity**
  the ratio between (i) the volume of water that a pervious material previously saturated with water will yield in specified hydraulic conditions, and (ii) the total volume of the material.

- **effective radius of a well**
  the equivalent radius of a hypothetical or theoretical well without any losses with the same axis as the real well and

having the same draw down for the same discharge.

- **effective rainfall**
the proportion of total rainfall which is useful or utilisable for productive purposes.

- **effective soil depth**
depth of soil up to which plant roots penetrate to draw water and plant nutrients.

- **effective water use**
consumptive use less evaporation from rainfall. It includes transpiration and evaporation from irrigation and transpiration from effective rainfall.

- **efficiency load range factor**
a term used in describing the characteristics of a water wheel, as an index of the manner in which the efficiency of operation of the wheel is maintained over a varying range of load.

- **efficiency of irrigation**
the fraction of the water diverted from a river or other source that is consumed by the crop, expressed as per cent.

- **efficiency of water application**
the fraction of the water delivered to the farm that is stored in the root zone for use by the crop, expressed as per cent.

- **efficient cropping zone**
it is a cropping zone in which-relative yield index is 200 per cent and relative spread indices vary from 90-200 per cent.

- **effluent stream, or gaining stream**
a stream fed by ground water.

- **effluent weir**
a weir at the outflow end of a sedimentation basin or other hydraulic structure.

- **ejector**
a device for moving a fluid or solid by entraining it in a high velocity steam, air, or water jet.

- **ejector pump**
a pump that consists of a centrifugal pump, which is located at the top of the well, and a hydraulic ejector near the water surface.

- **electric drill**
a drill having either a rotary motion provided by an electro-magnetic motor, or a percussion or reciprocating action provided by solenoids.

- **electric shock**
  a violent concussion that occurs when electrical current flows through the body.
- **electric tank heater**
  a device for heating liquids, consisting of an electric resistance coil which is introduced into the tank which holds the liquid.
- **electrical circuit**
  a pathway for the flow of electrons.
- **electrical conductivity**
  the relative ability of a material to allow electrical current flow.
- **electrical resistance moisture meter**
  a device used to determine the moisture content of grain by measuring the electrical resistances of grain sample as it is fed between two steel rolls which act as electrodes. One of the rolls is motor driven. The resistance is calibrated so it can be read in moisture content.

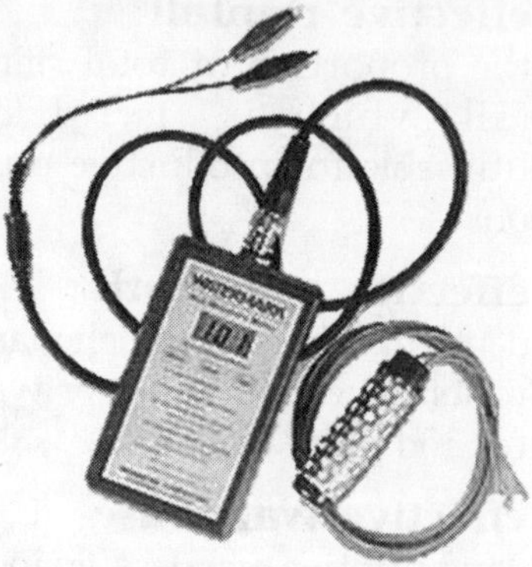

- **electricity**
  the flow of electrons from atom to atom in a conductor.
- **electro dialysis**
  an electro-reclamation process which increases the rate of percolation and opens up the soil.
- **electrodes**
  metals, which are dissimilar used in a chemical cell for producing electricity.
- **electrolyte**
  a chemical solution containing ions.
- **electromagnet**
  a soft iron core surrounded by a coil of wire that temporarily becomes a magnet when electrical current flows through the wire.

- **electromotive force**
the electrical force or pressure that causes or tends to cause current to flow in a circuit. It is commonly measured in volts.

- **electron**
a negatively charged atomic particle orbiting around an atom's nucleus.

- **electron theory**
theory stating that electricity is the flow of negatively charged particles (electrons) and that flow is from negative to positive.

- **electronics**
the use of semiconductors to manipulate electrical signals.

- **element**
substances that cannot be reduced into simpler substances using ordinary chemical means.

- **elevated ditch**
an earth-fill crossing, constructed to specification similar to those for earth-fill dams, placed across canyons or other large depression to provide canal or lateral crossings at normal grade, as a substitute for flumes or siphons.

- **elevated storage**
in any distribution system, storage of water in a tank supported on a tower.

- **elevated tank**
a tank used for storage purposes in a water distribution system, which is raised above the surface of the ground and is supported by posts or columns.

- **elevator**
a device on a drilling ring used in lifting pipe or casing.

- **elevator pump**
an endless band with buckets attached, running over two drums for draining shallow ground.

- **eliminator**
a type of moisture separator, in which suspended moisture is separated from air or other products by passing it through a vessel containing corrugated ribs and diaphragms, which collect the moisture and direct it to where it can be removed.

- **elutriation method**
a type of mechanical analysis of soil where the separation of different soil fractions is effected by washing out particles in a rising current of

water (particles whose setting velocity is equal to or less than the velocity of the rising water will be taken away by the current).

- **enclosure**
  the housing or covering of an electric motor.
- **end bell**
  the removable end-coverings of an electric motor.
- **energy**
  the capacity to do work.
- **engine**
  a device that changes heat energy to mechanical energy.
- **engine coolant**
  a mixture of anti-freeze and water used in liquid cooling systems to transfer heat from engine components to the surrounding air.
- **engine cooling system**
  system designed to maintain the optimal temperature of an internal combustion engine.
- **engine displacement**
  the total swept volume of an engine's cylinders as the engine completes one stroke.
- **engine timing**
  proper sequencing of the intake, compression, power and exhaust events of an internal combustion engine.
- **engineered grade**
  plywood grade based on approved uses for the material.
- **equal gabled roof**
  a gable roof with the same slope on both sides of the roof.
- **equipment grounding**
  connecting conductive equipment and surfaces that are not intended to carry electrical current to the earth.
- **equipment grounding conductor**
  a conductor used for connecting conductive, non-current-carrying parts of equipment, raceways, and other enclosures back to the system grounding electrode.
- **equivalent simple circuit**
  a circuit in which all series and parallel loads are combined, thereby forming a single load having the same resistance as the original circuit.
- **exhaust**
  1. the procedure of removing spent combustion products from the engine combustion chamber.
  2. the final stroke of the four-stroke engine cycle.

- **exhaust system**
auxiliary engine system that removes exhaust gases and particles from the combustion chamber.

- **f**
the suffix f is used for soil that is thought to be permanently frozen.

- **face**
the approximately vertical surface extending upward from the floor of a pit to the level at which drilling is being done.

- **factorial experimentation**
a method of experimentation where, in place of the classical method of varying only one experimental factor at a time, a comprehensive study is made of the number of experimental factors at the same time.

- **factors of soil formation**
a factor of soil formation is an agent, force, condition, or relationship, or combination of these, which influences, has influenced, or may influence a parent material of a soil, with the potential of changing it.

- **fahrenheit scale**
a scale in which the normal freezing point of water is 32° and the normal boiling point of water is 212°. Since there are 100° Celsius degree and 180 Fahrenheit degree between the freezing and boiling point of water, 5 celcius degree equal 9 Fahrenheit degrees. Formula can be written as: C = (F-32) 5/9

- **fair lead**
a device of sheaves or rollers to line up cable so that it will wind smoothly on to a drum.

- **fall increaser**
a device connected with a draft tube to increase the discharge by reducing the tail water pressure.

- **fall, drop**
a structure designed to secure the lowering of the water surface in a channel in a short distance, and the safe destruction of the liberated surplus energy.

- **falling crest weir**
a weir having falling shutters fitted along its crest which may be laid down to pass floods.

- **falling drop technique**
a technique that provides a measure of the stability of structural units to raindrop impact.

- **fallow**
  the practice of allowing crop land to lie idle during a growing season to build up the soil moisture and fertility content so that a better crop can be produced the following year.

- **fallow land**
  land left unfarmed for one or more growing seasons to kill weeds, make the soil richer.

- **fallow systems**
  the sequence of crop years with fallow years. Extensive fallow systems are shifting cultivation systems. Intensive fallow systems are bush fallow and grass fallow systems.

- **fallowing**
  keeping the land free of a crop or weeds for a period of time.

- **false set**
  setting of cement which rapidly develops a crust, sufficient to indicate an initial set, but which is not hardened in the interior and on reworking may set normally.

- **fan**
  accumulation of debris brought down by a stream on a steep gradient and deposited on a gently sloping plane in the shape of a fan.

- **fanglomerate**
  heterogeneous materials originally deposited as an alluvial fan and subsequently cemented into a solid rock.

- **fanning mill**
  a type of seed cleaning machine using screens or sieves and an air blast, resulting in a separation based on size, shape and specific gravity.

- **fanning plume**
  a smoke plume that occurs in stable air, when the gases quickly reach their highest level and then travel horizontally, remaining visible for a long distance — consequently, with little dilution. Fanning may thus produce a concentrated pollution on distant hills or tall buildings.

- **farad (F)**
  unit of measurement of capacitance.

- **farm**
  all the land which is directly farmed by one person, either by his own labour or with the assistance of his household or hired employees.

- **farm budget analysis**
  analysis of the farm's income and expenditure.

- **farm capital**
the value of all owned farm land, farm buildings excluding those meant for residential purposes, livestock, implements and tools owned by the operator.

- **farm drainage**
the removal, or exclusion, of excess water from farm land by open or closed drains and pumping plants.

- **farm duty of water**
the seasonal quantity of water delivered to individual farm units under an irrigation project. Such duty may be expressed in terms of acre-feet per acre, total depth of water over a unit area, or number of acres served per unit of flow.

- **farm elevator**
a grain storage building in which is installed a permanent mechanical device for moving grains. Also a mechanical device for moving agricultural products.

- **farm forestry**
farm forestry is a process in which trees are grown specifically for fuel, food and for a variety of other resources on a farm. It is a 'three dimensional farming' and, provides an alternative to monoculture grain farming. The three dimensions are defined as the trees themselves, the harvest from the trees which is used to feed livestock and the animals and the animal products.

- **farm inventory**
an itemised list of the land, buildings, equipment, livestock and supplies with their values at the date when the inventory is taken.

- **farm irrigation efficiency**
the ratio expressed in percentage of irrigation water available for crop production to that delivered at the farm head gate.

- **farm irrigation preparations**
the preparation of land for irrigation including (1) the selection of a suitable layout, (2) the provision of supply and drainage channels, to enable the efficient delivery and removal of water, (3) the

grading of the land, and formation to banks and subsidiary channels to facilitate the application of water.

- **farm irrigation procedures**
  procedures followed or to be followed on a particular farm in irrigated agriculture including (1) method of delivering water to field units, (2) frequency of irrigation, (3) depth of water applied for irrigation.

- **farm irrigation requirement**
  it depends on the irrigation need of individual crop, their area and the losses in farm water distribution systems, mainly through seepage.

- **farm irrigation structures**
  devices, pieces of equipment and small hydraulic structures used by the individual irrigators to divert water from the supply ditch to his head ditch and to convey it to the several parts of his farm including control, regulation and measurement of water.

- **farm losses**
  losses of water on the farm due to uneven distribution, poor handling, evaporation and percolation into the subsoil due to over irrigation.

- **farm management**
  the branch of agricultural economics which deals with the business principles and practices of farming with an object of obtaining the maximum possible return from the farm as a unit under a sound farming programme.

- **farm mechanics**
  the unspecialised, mechanical activities performed on the farm and in the home.

- **farm peaks**
  the amount of crop a farmer is permitted to grow.

- **farm planning**
  a process involving many decisions to be taken in respect of the kinds of crops to grow, rotations, mixtures, soil and water conservation practices to be followed and buildings, bullocks, machinery purchase etc.

- **farm processing**
  any method or treatment used to prepare farm products for

use or preservation is called farm processing.

- **farm sprinkler method**
a method exclusively for a given area or farm unit on which sprinkler irrigation is the primary method of water application.

- **farm waterway**
a natural or constructed waterway which is entirely within an individual farm boundary.

- **farmer**
one who cultivates the land.

- **farming systems**
farming systems represent an appropriate combination of farm enterprises viz., cropping systems, livestock, fisheries, forestry, poultry and the means available to the farmer to raise them for profitability. It interacts adequately with environment without dislocating the ecological and socio-economic balance on one hand and attempts to meet the national goals on the other.

- **farmyard**
that part of the farmstead that includes the area not occupied by the buildings.

- **farmyard manure spreader**
a machine to carry farmyard manure and spread it in a regulated quantity.

- **fascia plumb cut**
a cut at the bottom end of a rafter to which a fascia board is attached.

- **fascine**
a bundle of brushwood usually round, bound together with wire, etc.

- **fault dam spring**
a spring above a reverse fault, the depressed bedrock block being covered with a thick deposit of alluvium and the raised block opposing a barrier to the flow of ground water.

- **feeder canvas**
any wide canvas belt fitted with salts and used to elevate materials, usually cut grain and straw, to the threshing or other processing mechanism.

- **feeder circuit**
an electrical circuit that originates at a service entrance panel and supplies electrical power to a sub-panel.

- **fence line feed bunk**
a trough like structure which is designed for feeding livestock along a fence.

- **fencing**
a device that fixes the boundary of the farm and helps to stop the encroachment of the land and protection from wild animals and cattle.

- **fermentation layer**
in soils, a layer of partly decomposed litter lying below the freshly fallen organic material and above plant remains that have been fully decomposed to form humus.

- **ferrous**
containing iron (Fe).

- **ferrugenous soil**
soil containing large amounts of iron minerals especially limonite and hematite, characterised by red colour or shades of yellow and brown.

- **fertigation**
application of fertilisers with irrigation water.

- **fertiliser ratio**
the fertiliser ratio designates the relative proportion of three major plant nutrients, keeping the percentage of nitrogen as one of the ratio.

- **fertilisation**
the process of adding plant food to the soil, for purpose of plant growth.

- **fertiliser**
any natural or manufactured material, dry or liquid, added to the soil in order to supply one or more plant nutrients. The term is generally applied to commercially manufactured materials other than lime or gypsum.

- **fertiliser application**
the use of materials containing plant nutrients on soil.

- **fertiliser applicator**
equipment connected with the lateral line of a sprinkler sys-

tem for injecting water soluble fertiliser into the water to be sprinkled.

- **fertiliser attachment**
device attached to an implement or machinery to facilitate distribution of fertiliser in or upon the soil.

- **fertiliser boot, split band**
a fertiliser placement device which allows placement of the fertiliser in bands on either side of the seed, usually above the seed level.

- **fertiliser broadcaster**
a fertiliser distributor with a spreading width substantially greater than the width of the machine.

- **fertiliser distributor**
equipment used to apply fertiliser in the desired way, including broadcasting, placement, deep application, etc.

- **fertiliser drill**
a machine to deposit fertiliser in soil at regulated and selected rates and at predetermined depth.

- **fertiliser gun**
a machine to throw a jet of fertiliser by mechanical, pneumatic or other means, usually fitted to the side of the machine.

- **fertiliser mixture**
a product obtained by physically mixing different fertilisers so as to contain more than one of the three major nutrients, viz., nitrogen, phosphorus and potassium.

- **fertiliser ratio**
the fertiliser ratio designates the relative proportion of three major plant nutrients, keeping the percentage of nitrogen as one of the ratio.

- **fertiliser, starter**
a relatively small application of fertiliser applied with or near the seed for accelerating early growth of the crop.

- **fertility (of soil)**
the quality that enables a soil to provide the proper compounds, in the proper amounts and in the proper balance, for the growth of specified plants when other factors, such as light, temperature, and the physical condition of the soil, are favourable.

- **fibre saturation point**
moisture condition in which all water, except that in the cell walls, has been removed from wood.

- **fibrillation**
rapid, uncontrolled contractions of portions of the heart

muscle that cause irregular heartbeats.

- **field border plantings**
  vegetation established on border of fields in order to conserve soil and provide cover and food for birds etc.

- **field chopping**
  the harvesting of forage, corn and other green crops, and straw by a machine which chops the material in the field and delivers it into a wagon or truck. Forage may be chopped directly as standing material or picked up from windrows.

- **field cultivator**
  an implement which has several sweeps or shovels so mounted as to provide subsurface tillage. The standards which carry the sweeps are staggered to permit trash to flow around and between them.

- **field drain**
  a land drain.

- **field head capacity curve for pump**
  a curve showing how much water a given pump will deliver with a given head.

- **field implement**
  any type of machine designed for use in the field for the production of agricultural crops as compared to those used as stationary installations.

- **field irrigation efficiency**
  overall efficiency in the irrigation of a given field or plot of land and consists in examining together moisture distribution factors, moisture storage capacity factor and field water application efficiency.

- **field pit**
  a pit which is dug in or near fields of potatoes or cabbages and in which these products are placed for storage, covered by soil or layers of soil and straw to protect them from the winter cold.

- **field storage**
  the storage of produce in trenches and pits or by mounding on the surface of the ground and in above ground horizontal silos.

- **field strip cropping**
  a system of strip cropping in which crops are grown in parallel strips across the general slope of the land rather than on the contour. Strips of close growing crops are alternated with those of wide spaced cultivated crops.

▪ **field test**
an experiment conducted under field conditions. Ordinarily less subject to control than a formal experiment and may be less precise.

▪ **field-effect transistor (FET)**
a voltage controlled transistor.

▪ **fields**
subdivisions of a farm which form areas on which specific crops are grown, or for pasture purposes.

▪ **file**
a steel instrument, having its surface covered with sharp edged furrows or teeth, which is used for sharpening, abrading or smoothing other substances, such as metals, wood or stone.

▪ **filter**
a device used for removing small particles from a fluid.

▪ **filter crop**
close growing crops planted across the slope to retard run-off and cause deposition of the soil load.

▪ **filter efficiency**
the operating results from a filter as measured by various criteria such as percentage reduction in suspended matter, total solids, biochemical oxygen demand, bacteria, colour, etc.

▪ **filter point wells**
wells generally drilled in delta and coastal areas where the substrata is soft and sandy.

▪ **filter strip**
a strip of permanent vegetation of sufficient width and vegetative density grown above farm ponds, diversion terraces and other structures in order to retard flow of run-off water and encourage deposit of transported soil; thereby avoiding silting of structure or reservoir below.

▪ **filtered water reservoir**
a reservoir for the storage of filtered water of sufficient capacity to prevent the necessity of frequent variations in the rate of filtration, with variations in demands.

▪ **fin**
steel sheeting wall projecting from the main coffer dam structure.

▪ **final crop**
that portion of the stand which is intended to retain to the end of the rotation.

- **final yield**
all the material that counts against the prescribed yield derived from the principal fillings in a regular forest.

- **finger feed**
as applied to fertiliser distributors, it refers to a type of metering device which consists of a flat rotating disk having projections on its periphery which transfers fertiliser from the hopper to the delivery tube. It is commonly called a STAR WHEEL feed.

- **fire curing**
a process of curing tobacco specially for tobacco used for chewing purposes.

- **fire harrowing**
burning of organic material by dragging harrows containing lighted accumulated material over the ground surface; used where there has been a poor initial burn.

- **fire plough**
a heavy duty, usually specialised machine, either of the share or disk type, which is designed solely for abusive work in the woods and is used with either horses or tractors to construct fire breaks and fire lines.

- **firing order**
the sequence of power strokes in the various cylinders of a multi-cylinder engine.

- **Fischer method**
an indirect method of determining moisture content of a product by the use of a chemical.

- **fish transplanting**
transfer of migratory fish from spawning grounds by the construction of a high dam to basins that drain into rivers at downstream locations, for either spawning in them, if conditions permit, or to place them in holding ponds to spawn, where after they are released in the streams to begin their downstream migration.

- **fittings**
coupling joints, valves, tees, etc., for pipes which are used for various purposes; that is, sprinkler irrigation system; dairy apparatus, high pressure fittings for lubrication of wheels or other moving parts.

- **fixed capacitor**
a capacitor having a set, non-adjustable capacitance value.

- **fixed dam**
a type of dam with no provisions for moving the water

face. A fixed dam may have a movable crest in the form of flashboards, gates, etc.

- **fixed displacement pump**
  a pump in which the volume of the pumping chamber cannot be changed while the pump is in operation.
- **fixed head, sprinkler**
  a type of sprinkler fixed in position, having no moving parts and capable of dispersing water in a full part circle or square patterns.
- **fixed resistor**
  a resistor that has a set, non-adjustable resistance value.
- **flail**
  a wooden handled implement of many designs at the end of which it fastened a free swinging metal or wooden piece called a swingle which is used to thresh the grain by hand from harvested plants by beating the seeds free from the straw to thresh the grain from the stem by the use of the flail.
- **flat belt**
  a thin, wide belt made up of leather or several piles of canvas impregnated with rubber.
- **flat drop seed plate**
  a type of plate used in cell type seed metering devices which has openings around the periphery of the plate large enough to allow a single seed to lie flat.
- **flat furrow ploughing**
  ploughing in such a manner that the furrow slices are inverted and lie flat on the bottom of the adjacent furrow.
- **flat head**
  type of cylinder head that does not house the valves.
- **flexible coupling**
  a shaft coupling used to connect two shafts in which perfectly rigid alignment is impossible; the drive is commonly transmitted from one shaft to another through a resilient joint such as a steel spring, a rubber disk or bushings.
- **flexible harrow**
  a spike-tooth harrow with hinged links between the tooth bars which permit it to be rolled up on one steel bar for transport.
- **float or smoother**
  a framework of plants or steel angles, plates, etc., for drawing over the soil, to compact

and smoothen its surface, to improve its condition or to crush clods.

- **floating weeds**
weeds which either are not anchored in the soil and float freely on the water or are attached to the sides or bottom of the ditch and float only in a fixed area.

- **flocculated soil**
a soil in which the solution is more or less acid, but which shows no pronounced migration of clay.

- **flocculation**
the grouping of solids in water, resulting in floes, often wrongly thought to be the same as coagulation.

- **flood**
a relatively high flow or stage in a river, markedly higher than the usual; also the inundation of low land which may result there from.

- **flood benefits**
benefits resulting from the construction of reservoirs, levees, channel improvements, or other flood protective works, as estimated in terms of drainage eliminated or other advantageous effects of the proposed works.

- **flood control benefits**
direct benefits based on reduction in flood damage to land and other property in terms of cost of restoration to pre-flood condition or reduction in its value and reduced damages to crops in terms of market value adjusted by replanting possibilities and production costs not incurred; and indirect benefits based on higher grade use of land formerly flooded in terms of increased earnings, and reduced interruption of business, industry and commerce, traffic, communications and other activities both within and outside the area subject to flooding in terms of net loss of income or added operating costs and cost of evacuation, re-occupation, temporary quarters, emergency flood protective work, and relief and care of flood victims.

- **flooding soil**
1. a condition in which there is free water present at the surface of the soil.
2.water from field channels is allowed to flood the fields in irrigation by flooding.

- **floor plan**
a drawing depicting the overall layout of a building.

- **flow**
  a stream of water; movement of water, silt, sand, etc.; discharge; total quantity carried by a stream.
- **flow irrigation**
  the system of irrigation of lands by gravity where the irrigation water is available to a higher level than the irrigated land.
- **fluid fertiliser**
  liquid fertilisers like solutions, suspensions and slurries.
- **fluorescent light**
  a type of bulb consisting of a glass tube filled with a gas and having a filament in each end.
- **flush**
  to irrigate a field with just enough water to soften the crust. To wash a surface with a large volume of water applied for only a short time.
- **flush irrigation**
  irrigation in which water flows over the surface of the ground in a continuous sheet of shallow depth, spreading slowly and soaking in.
- **fluvial**
  pertaining to streams.
- **foliar fertiliser (foliar feed)**
  a product containing major, secondary and micro-nutrients, as required, designed for application to, and nutrient uptake by the foliage of the crop; usually applied at a low concentration.
- **footing**
  the projecting base under a wall or column that spreads weight and helps prevent settling.
- **force**
  a pull or push that tends to start, stop or change the direction of motion of an object.
- **forced draught fan**
  a fan that blows air into a furnace. It therefore handles only cool air and is subject to much less severe working conditions than an induced-draught fan, but both are needed in some installations.
- **forecast**
  a statement of anticipated meteorological conditions for a specific place (or area) and period of time.
- **forest soil**
  great variation, fine in texture surface layer consists of well decomposed humus and mineral soils, strongly acidic in reaction, high base exchange capacity is due to high organic matter content, water soluble Mn is present, suitable for forest trees.

- **forward-biased**
a condition in which the anode of an electronic device is more positive than the cathode.

- **fouling**
a gelatinous slimy accumulation on the waterway of a conduit, resulting from the activity of organisms in waters. Fouling may be found on concrete, masonry, and metal surface, but tuberculation is found only on metal surfaces and is harder to remove than fouling.

- **foundation tunnels**
tunnels driven in the rock below the base of a dam and also into the abutments at a higher elevation for grouting; in order to provide access for foundation grouting operations, they also provide a means of access to the foundation for any future treatment in case unsatisfactory sub-surface conditions develop after the structure is completed and the reservoir filled.

- **foundation wall**
structural supporting component from the footing to the above grade wall.

- **frame**
the primary housing of an electric motor.

- **frame number (FR)**
number found on a motor nameplate that, when divided by 16, gives the number of inches from the centre line of the motor shaft to the bottom of the motor mounting.

- **framed dam**
a low fixed dam generally built of timber framed to form a water face, supported by struts.

- **frameless type tractor**
a method of tractor construction which involves the use of two distinct castings one serving as the engine crank case, the other as the transmission housing, the two being bolted together to function as a frame.

- **free electrons**
electrons not tightly held in orbit around the nucleus of an atom.

- **free ground water**
ground water in the interconnected interstices in the zone of saturation; it extends down to the first impervious barrier, and moves under the influence of gravity in the direction of the slope of the water table.

- **free humus**
that fraction of humus which can be extracted by ammonium oxalate.

- **freeze drying**
removing moisture with cold.

- **French drain**
a sloping trench back—filled with coarse stones, which drains away surface and subsurface water, usually with the help of a land drain.

- **friction**
resistance to motion of two objects or surfaces that are in contact.

- **friction power (Pf)**
1. the power an engine consumed in overcoming friction.
2. the difference between an engine's indicated power and brake power.

- **frow**
a cleaving tool, with a wedge-shaped blade which is used by coopers in splitting staves for casks, etc., and in making shingles and clapboards. The handle is at right angles to the back of the blade and is held in the left hand, while the mallet which drives the edge into the log is held in the right hand.

- **fuel cell**
machine for producing electricity from the reaction of chemicals supplied to the fuel cell from an outside source.

- **fuel combustion meter**
engine analysis instrument that is designed to determine the percentage of a fuel gas actually burned by the engine.

- **fuel injectors**
three types: single hole fuel valves for pre-combustion chambers; multi-hole nozzles for combustion chambers of direct injection types; and the pintle type fuel injectors for turbulence chambers.

- **fuel system**
engine auxiliary system that is designed to meter and deliver clean fuel to the combustion chamber.

▪ **full reservoir level**
the highest reservoir level that can be maintained without spillway discharge or without passing water downstream through sluiceways. It does not include any depth of surcharge.

▪ **full-flow pressure**
the pressure required to completely open a hydraulic pressure relief valve.

▪ **full-wave rectification**
the process of changing both halves of an AC sine wave into DC electricity.

▪ **fully integrated operation**
integrated operation as applied to all the beneficial purposes for which water or water derived services are available and might be used within the area of service.

▪ **fully penetrating artesian well**
an artesian well which reaches the bottom of the confined aquifer.

▪ **fuel injectors**
three types: single hole fuel valves for pre-combustion chambers; multi-hole nozzles for combustion chambers of direct injection types; and the pintle type fuel injectors for turbulence chambers.

▪ **funnel flow**
flow of grain from a bin in which the volume of grain directly above the opening flows out first. Then the grain falls in from the top and the rest of the bin empties from the top down. If the opening is in the side of the bin a narrow column along the bin edge flows out and then the grain empties from the top down.

▪ **furrow planting**
planting in the bottom of furrows.

▪ **fuse**
a tool commonly used for protecting electrical circuits and loads from over-current conditions.

▪ **g**
a suffix symbol for strong gleying; used with a horizon designation to indicate intense reduction of iron during soil development, or reducing conditions due to stagnant water, as evidenced by base colours that approach natural, with or without mottles.

▪ **gauge**
a device for indicating or registering magnitude or position

in specific units, for example, the elevation of a water surface, the velocity of flowing water, the pressure of water, the amount of intensity of precipitation, the depth of snowfall, etc.

- **gauge rod**
a graduated measuring device which is designed to be partially immersed into a liquid and to be read at the point where the liquid surface crosses to rod.

- **gauge well**
a chamber or a compartment with closed sides and bottom, provided with an inlet or inlets communicating with the river or canal. Its purpose is to dampen waves or surges in the parent stream while permitting the water level in the well to rise and fall with the major fluctuations of the level in the river or canal.

- **galvanising**
the procedure of applying a thin coating of zinc to the surface of a metal in order to slow down oxidation.

- **gambrel roof**
a kind of roof where two different roof angles are used on each side.

- **gang**
a combination of implements or tools used simultaneously in the cultivation of soil.

- **gang of wells**
a group of wells from which water is drawn by a single pump or other water lifting device.

- **gang plough**
a plough with two or more bottoms attached to its frame.

- **garden agriculture**
crops grown on land more or less adjacent to settlements; usually vegetables etc., rather than staples, dependent on organic wastes from settlement for nutrient inputs.

- **garden hoe**
a hand tool which consists of a long wooden handle at one end of which a steel cutting blade is attached approximately at right angles. It is used in weeding, in cultivating, in making small furrows for garden seeds, etc.

■ **garden tractor**
any gasoline-engine powered tractor, usually not over five horsepower, which is used to plough and cultivate gardens. It may also furnish power for sprayers, pumps, weed bars, snow plough, carts, etc.

■ **gas, lift flowing well**
a well in which the ground water tapped is not under sufficient hydrostatic pressure to be lifted above the land surface without the introduction of gas but is caused to rise above the surface because of the buoyancy of natural gas mingled with water.

■ **gaseous fertiliser distributor**
a machine for applying mineral fertiliser in gaseous form which injects the fertiliser under pressure into the ground.

■ **gate chamber of sluice well**
a structure located within the abutment main body of the dam or the (3am embankment within which the regulating gates or emergency gates are installed. The lower portion of the chamber provides an anchorage for the gates and liner, while the upper portion furnishes a working space for the removal of the gate leaf and other parts for repair, and for the installation of the necessary control mechanisms. The gate chamber is usually located upstream of the axis of the dam.

■ **gear type pump**
a pump using a pair of matching gears that rotate; meshing of the gears forces oil (or other liquid) from between the teeth through the pump outlet.

■ **general shear failure**
failure in which the ultimate strength of the soil is mobilised along the entire potential surface of sliding before the structure supported by the soil is impaired by excessive movement.

■ **generalised soil map**
small-scale maps made to bring out the contrasts within large areas by generalisation of more detailed maps. They vary from soil association maps of a country on a scale of 1 inch equals 1 mile (1: 63,

360) to maps of larger regions showing associations dominated by one or more Great Soil Groups.

- **genesis soil**
mode of origin of the soil; refers particularly to the processes responsible for the development of the solum from the unconsolidated parent malarial.

- **genetic system of soil classification**
any system of soil classification in which the genesis or the history of the evolution of the soil constitutes the basis. It recognises the fact that distinguishing characteristics, physical, chemical and biological, in the successive horizons of soil profiles drawn from different areas of the world arise out of varied degree of interaction of soil forming factors.

- **geographic information system (GIS)**
a computerised field mapping process where fields are divided into grids and then mapped for physical attributes per grid.

- **geohydrology**
the branch of hydrology that deals with ground water, its occurrence and movements and its replenishment and depletion.

- **germanium**
a chemical element that has four electrons in the valence ring.

- **ginaca machine**
a machine used in the canning of pineapples which automatically cuts a cylinder from the central portion of the fruit, removes the shell, cuts off the shell at each end of the fruit, removes the core, and scrapes out of the edible flesh remaining on the shell.

- **ginning**
the process of removing the seeds from cotton.

- **girard turbine**
an impulse turbine with buckets so shaped that on leaving the wheel the water has a relative velocity in direction opposite to that of motion of the wheel.

- **glacio alluvial deposit**
a deposit laid down by a stream originating in a glacier. Such deposits, while consisting of glacial drift, are usually well sorted by the running water, which carries the finer materials a greater dis-

tance from the glacier before dropping them.

- **gley**
a portion of a soil altered by the water which persists within it; the iron has been reduced to the ferrous state, providing characteristics bluish or pale green colours throughout the soil mass or only in spots or streaks. It results in destruction of the structure, existence of asphyxiating conditions and impoverishment by leaching and reduced biological activity.

- **gleying (rice culture)**
maintaining soil layer under reduced condition. This may occur due to water table being nearer to the surface or maintaining water on the soil surface during most of the rice growing season.

- **global positioning system (GPS)**
a system of high altitude satellites orbiting the earth and having the capability to provide exact geographic locations on the earth's surface.

- **governing system**
the engine auxiliary system that maintains a constant engine speed under varying load conditions by controlling fuel flow to the engine.

- **governor**
the primary sensing and control mechanism of an engine's governor system.

- **grab**
a two- or three pronged tool, with barbs on the inside of the prongs, for recovering broken ropes from bore hole.

- **grade stamp**
a stamp that indicates the quality of lumber or manufactured wood products.

- **grades of hardness**
a method for designating the lead hardness of drawing pencils.

- **grafting**
joining two parts of a plant together so that they grow together as one.

- **grain**
1. a unit of weight equal to 0.0648 part of a gram, 0.000143 part of an avoirdupois pound, and 0.04167 part of a pennyweight.
2. the direction of the fibres with respect to the longitudinal axis of the tree, log, the edges of a sawed piece of wood or laminated wood.
3. also a general class of farm crops such as wheat, oats, corn, etc.

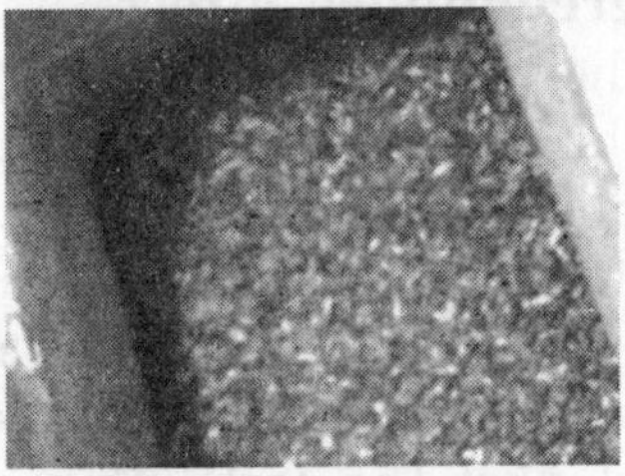

- **grain cleaner**
a machine to remove foreign matter from .grain mass.

- **grain drying**
drying the harvested grain usually in sun to reduce the grain moisture content to an acceptable level. Drying process is basically the transfer of heat by converting the water in grain to a vapour and transferring it to the atmosphere.

- **grain header**
a grain harvesting device which cuts off the heads of the grain and elevates them to a wagon. It is now largely replaced by the combine.

- **grain pan**
also, a metal pan on a grain thrasher or combine onto which grain falls after being separated by the straw rack. The grain is conveyed to the head end of the cleaning shoe either by an oscillating motion, or by a mechanical drag conveyor.

- **grain separator**
a machine to remove impurities from grain or other seeds and to sort them into two or more fractions.

- **granary**
a store house or repository for grain after it is thrashed or husked.

- **grape hoe**
a machine which is used for mechanical hoeing between grape or blueberry plants in the row.

- **grass seed attachment**
a device attached to a grain drill for the purpose of seeding grasses and legumes. It is so designed that it will place the seed to the rear of the grain drill either on top of the ground or in a band over the furrows left by the drill.

- **grass seeder**
an attachment used on grain drills for the purpose of distributing small grass and legume seeds uniformly.

- **grass stick**
a three to four ft. long stick attached to a grass board to further aid in turning the cut forage away from the uncut forage.

- **grass waterway**
  a strip of grass growing in the low area of a field where water can gather and cause erosion

- **gravitational moisture**
  water that moves freely in response to gravity and drains out of the soil.

- **gravity dam**
  it is customary to confine the term gravity dam to solid masonry or concrete dams and which resist their imposed loading by the weight of the structures themselves.

- **gravity irrigation**
  irrigation in which water flows and is distributed by gravity.

- **green manure**
  the biomass of a crop grown specifically for incorporation in the soil to improve fertility.

- **green water**
  water goes green in ponds in summer because of green *algae* which develop fast in sunlight. Their existence, however, may encourage other growths, including some that eat them.

- **greenhouse**
  a house covered with glass used for the protection and cultivation of plants.

- **greenhouse effect**
  the fact that the heat from the sun enters a greenhouse more easily than it can leave it. On the atmosphere as a whole, water vapour acts similarly, because clouds can blanket the earth at night, preventing frost. Carbon dioxide also has a greenhouse effect. It has therefore been claimed that the CO_2 in the air, if it were to rise much above its present level of 0.03 per cent, because of high consumption of fuels, could warm the earth by a greenhouse effect.

- **gross commanded area**
  the portion of the gross irrigable area which can be commanded by gravity irrigation.

- **gross cropped area**
  the total area covered with crops-during the year.

- **gross duty of water**
  the water requirement or duty, measured at the point where the water is diverted from its, source, including all canal or conduit losses, seepage losses, evaporation and transpiration losses in the system, and all waste occurring.

- **gross irrigable area**
the gross area less such areas as are excluded from the project by reasons of their being unsuitable for irrigation, either on account of the nature of the soil, or because of ground is too high to be irrigated by gravity flow or economically by pumps or other lifting devices.

- **gross irrigation requirement**
the total amount of water applied through irrigation including losses during conveyance, distribution and application of water in the fields etc.

- **ground water**
the water that occurs in the zone of saturation, from which wells and springs or open channels are fed.

- **grounded conductor**
circuit conductor intentionally connected to ground at the service entrance.

- **ground-fault**
an extremely dangerous electrical circuit condition in which a current-carrying wire makes electrical contact with a conductive object that is grounded.

- **ground-fault circuit interrupter (GFCI)**
an electrical apparatus intended to protect persons from electrical shock by de-energising the circuit almost instantaneously when a ground-fault occurs.

- **grounding**
direct or indirect electrical connection to the earth.

- **grounding electrode**
serves as the terminal or connection to earth for an electrical grounding system.

- **groundwater hydrology**
the branch of hydrology that treats of ground water; its occurrence and movements; its replenishment and depletion; the properties of rocks that control groundwater movement and storage; and the methods of investigation and utilisation of ground water.

- **groundwater irrigation**
irrigation by water derived from wells.

- **groundwater replenishment**
refilling or recouping of the depleted normal groundwater storage by natural processes or artificial methods.

■ **growth water**
water which is truly useable by the roots of plants.

■ **grub ax**
a tool which has two cutting edges for digging up roots; one a broad chisel-shaped point and the other a flat adze like blade.

■ **gully erosion**
process of erosion by which running water accumulates in narrow channels which go on deepening and widening as the process continues. Depth of channels or gullies may be as much as 1 to 30 metres and top widths 5 to 15 metres or more.

■ **gully, or gulley**
a small, elongated depression, usually eroded, in the land surface, usually dry except after a rainstorm; a channel or miniature valley cut by running water, but through which water commonly flows only during and immediately after heavy rains or during the melting of snow.

■ **gumbo**
a class of peculiar, fine-grained soils usually devoid of sand and rich in alkaline compounds. When saturated with water they become impervious and waxy or soapy in appearance and to the touch.

■ **guttation**
process by which plants expel liquid water from uninjured leaves in excess of transpiration

■ **gypsum board**
a manufactured construction material that is used primarily for interior walls. It is also called sheet rock or drywall.

■ **gyro sifter**
a machine in which grains of smaller size are separated by passing them over gyrating screens. Small size grains tend to get collected at the centre while the larger ones move towards the periphery and get separated.

■ **h**
a suffix symbol for illuvial humus; used for accumulations of decomposed illuvial organic matter, appearing as dark coatings on sand or silt particles, or as discrete dark pellets of silt size, are indicated by h. If used, this suffix follows the letter B or a subdivision of B, as Bh or B2h.

■ **h beam**
a steel beam with a cross-section in the shape of an H.

- **hackler**
a machine for removing the seed from the brushes.

- **half bog soil**
a great soil group of the intrazonal order and hydromorphic suborder consisting of soil with dark-brown or black peaty material over greyish and rust mottled mineral soil.

- **half shaft**
one of the two rear axles of a truck, car or tractor.

- **half shovel**
a cultivator shovel with a side wing on only one side, thus allowing cultivator shovel to pass through the soil close to the plant without danger of disturbance of the roots of the plant by the wing of the shovel.

- **half-wave rectification**
the process of changing only one half of an AC sine wave into DC electricity.

- **halomorphic soil**
a suborder of the intrazonal order consisting of saline and alkali soils developed under inadequate drainage in arid areas.

- **halophytes**
a plant which will live in a soil containing an appreciable amount of common salt or of other inorganic salts.

- **hammer mill**
a crusher consisting of a higher-speed rotating shaft with suspended hammers which repeatedly strike the rock and throw it against breaker plates, thereby successively breaking the stone to small particles.

- **hand cultivator**
a garden tool consisting of a handle or handles at one end of which are attached three or five curved rods or shovels which are used to break soil crust and to weed in gardens.

- **hand drill**
1. a drilling machine fitted to be worked by manual power.
2. a light drilling apparatus in which the drill is operated by means of a fiddle bow, a swing or geared brace or a ratchet.

- **hand hacksaw**
a hand powered tool for cutting metals.

- **hand harvest**
to pick or gather by hand as contrasted with mechanical harvesting by combine, corn picker, etc.

- **hand tool**
any tool or implement, cutting or otherwise, which is used manually as distinguished from a machine tool.

- **hand truck**
a piece of handling equipment used to carry a load. It usually has an L-shaped frame mounted on two wheels and a pair of handles separated by stretchers so that a man can easily guide and push the truck. Some trucks have four wheels.

- **hard cemented soil**
when soil is ploughed too dry, the normal packing of the soil undergoes shrinkage, cementation effects become large and the soil has a harsh consistency.

- **hardboard**
a manufactured wood material which is made by pressing wood fibre mats together under high temperature and pressure conditions.

- **hardening**
a heat treating process that increases the relative hardness of a metal.

- **hardness**
the relative capacity of a metal to resist indentation, scratching, or filing.

- **hardway or head grain**
the direction at right angles to both the rift and grain in which the stone does not split readily.

- **hardy crop**
plant which survives even when exposed to unfavourable conditions like prolonged droughts. Ragi (finger millet), moth bean and guar (cluster beans) are considered as hardy crops.

- **harp**
a tool consisting of a frame on which parallel wires are stretched for cutting the coagulum of milk for cheese making.

- **harrow**
a cultivating implement used primarily for breaking clods and pulverizing and smoothing the ploughed soil and sometimes for mulching, covering seed, and removing weeds.

- **harrow cart**
an attachment for a horse drawn harrow which consists of a seat mounted between two wheels for the convenience of the operator.

- **harrow plough**
a type of disk harrow with all disks mounted in a com-

mon axis, throwing the soil in one direction. There may be from 4 to 30 or more disks, which are usually larger in diameter than disks used in a conventional disk harrow.

- **harrow teeth**
the points of a harrow which are in contact with the soil.

- **harrow, disk**
a series of concave disks mounted on a central axis and usually operated at an oblique angle to the direction of travel through the soil.

- **harrowing**
working the soil with the help of a wooden implement called harrow.

- **harvester**
an agricultural implement, a further development of the reaper, in which the cut grain falling on the travelling canvas belts of a conveyor is gathered into sheaves, bound with string, and discharged clear of the machine.

- **harvester thresher**
an implement which includes the combined operations of harvesting and threshing grains.

- **harvesting**
the operation of cutting, picking, plucking or digging, or a combination of these for removing the useful part or economic part from the plants.

- **hatchery**
the place where eggs are hatched.

- **hatchet**
a small axe which is used for splitting or rough dressing timber.

- **hay**
hay is feed produced by dehydrating green forage to a moisture content of 15 per cent or less so that the biological processes do not proceed rapidly enough to build up heat.

- **hay baler**
a machine which is used for compressing loose hay into compact bales and securing them with twine or wire. There are two principal types, the stationary and the field baler. The field or pick-up baler is most extensively used. This baler

picks up hay from a windrow and the bale is made while the machine is in motion.

- **hay blower**
an attachment which is used with a hay chopper to blow the chopped hay into the hauling truck or into the storage.

- **hay chute**
an opening in a two-storey barn, to throw hay through to the downstairs, usually has a trap door or a protective railing enclosing it.

- **hay hook**
a steel hook sharpened at one end and with a hand grip on the other. It is inserted into the end or side of a bale of hay, assisting in easier and faster handling.

- **hay press**
an implement used to press loose hay into a bale.

- **hay rack**
a slotted wooden frame used to contain hay prior to its consumption by livestock. A wooden frame, sometimes called a basket, usually about 8 feet wide and 16 feet to 18 feet long, which is placed on the running gears of a wagon and used to haul loose or baled hay or bundles from field to barn stack, threshing machine, other storage, or to market.

- **hay rake**
a machine used for gathering cut hay and placing it in bunches and windrows. Any of several different kinds of rakes.

- **hay sling**
a device used for unloading hay from wagon or hay rack to the hay stack or mow.

- **haylage**
product resulting from ensiling forage with 35-50 per cent moisture under anaerobic conditions. Compared to hay, it greatly reduces the time involved in field curing and the associated field risk.

- **head capacity curve for a well**
a curve constructed by plotting data obtained by installing a test pump in a well which, indicates how much water well will yield at any given level of draw down.

- **head control structures**
temporary dams used to control ditch levels where desirable in irrigation.

- **head gate structure**
the portion of the head regulator which houses control gates.

- **head land**
while ploughing with a tractor a strip of un-ploughed

land is left at each end of the field for the tractor to turn, which is called head land.

- **head loss**
energy lost (converted into heat) as the result of friction, impact or eddies expressed as a 'head' per unit of flow.

- **head of a canal**
the point where the supply is diverted into a canal from a reservoir, river, or parent canal.

- **head of water**
the energy in water that enables it do mechanical work. The total head is the sum of three components: (1) elevation head—for example, the height above sea level; (2) pressure head, measured by a piezometer tube, is the height of the static column of water that can be carried by the pressure at that point; (3) velocity head, the energy in water due to its movement (its kinetic energy) is equal to $v^2/2g$, where v is the mean velocity of the water and g is the acceleration due to gravity. This kinetic energy is measured by a Pitot tube with the opening facing upstream.

- **head water erosion**
erosion which occurs in the upstream end of the valley of a stream, causing it to lengthen its course in such direction.

- **head water storage**
works designed to store or otherwise retain rain water, where at falls or as near thereto as practicable, to hold back the crest.

- **head works**
a general term applied to all the structures, devices, etc. located at the head or diversion point of a conduit or canal. The term used is practically synonymous with diversion works; an intake heading.

- **headrace**
a channel which diverts water to a waterwheel.

- **headwall**
a wall of stone, metal, concrete, or wood at the end of a culvert or drain to serve one or all of the following purposes; protect fill from scour or undermining, increase hydraulic efficiency of conduit, divert direction of flow, retard disjointing of short sectional pips, and to serve as a retaining wall.

- **headward erosion**
the erosion of a stream bed by the development of scour

at the bottom of a vertical cliff or waterfall over which the flow of the stream plunges, causing undermining of the cliff and its movement upstream, so that the deepened section extends upstream.

- **headwater**
  water level prevailing in the supply channel.

- **hearting**
  the portion of the main body of a dam not affected by atmosphere.

- **heartwood**
  the wood at the centre of a tree or branch.

- **heat pump**
  a refrigeration machine installed to make use of the heat discharged from the condenser instead of the heat absorbed by the evaporator. It pumps heat from a low temperature source for use at a higher temperature.

- **heat recovery wheel**
  a large wheel that slowly turns through hot outgoing air while its other half is turning through cool incoming air. In this way much of the heat from the outgoing air is transferred to the incoming air, as in the rotary regenerative air heater.

- **heat sink**
  a device for absorbing heat from one medium and transferring it to another. The diodes in a-c generators are usually mounted in heat sinks which remove the heat from the diodes and thus prevent them from overheating.

- **heat treating**
  the process of changing the properties of a metal by controlling the heating and cooling process.

- **heating values**
  the heat given off by a fuel when it is burned. Usually expressed in Btu per unit weight or unit volume.

- **heave**
  upward movement of soil caused by expansion or displacement resulting from phenomena, such as moisture absorption, removal of overburden, driving of piles, and frost action.

- **heaving**
  horizontal or lateral displacement of soil formation surrounding the well hole while pumping during digging which occurs when the differences in head between the inside and outside water levels are sufficient to allow the sand

and gravel in the excavation to blow up.

- **heavy (textured)**
applied to soils of fine texture in which, clay predominates, with dense structure and firm to compact consistency. The term is also applied to soils containing a somewhat higher proportion of the finer separates than is typical of that textural class (as a heavy sandy loam).

- **heavy soil**
one that contains relatively high proportions of clay and therefore is not easily worked. It is heavy work for a team to pull a plough through it. Hence clayey soils are termed heavy ones.

- **heel**
the line joining the upstream slope and face of a dam with the valley floor or upstream blanket.

- **heel cutting**
a shield-shaped cut made about halfway through the wood around the leaf and the axial bud

- **heel line**
the line joining the upstream slope or face of a dam with the valley floor or upstream blanket.

- **height of bucket lip**
the distance from the lowest concave surface of the bucket to the top of the lip.

- **height of dam**
the difference in elevation between the roadway, walkway, or spillway crest and the lowest part of the excavated foundation along the axis of a fixed dam.

- **Henry (H)**
the unit of measurement for inductance. 1 Henry is the inductance required to induce 1 volt in a coil when the current changes its rate of flow by 1 ampere per second.

- **herb**
plant having no persistent parts above the ground, unlike a shrub or tree.

- **herringbone system**
a system in which the main drain and the laterals have a herringbone arrangement.

- **hertz (Hz)**
indicator of the number of cycles per second for AC electricity. For instance, electricity in the US is generated at 60 Hz.

- **heterometry**
the character of a material which is composed of grains of different sizes.

- **hexa-chrome**
chromium in the form of chromium—CrO_4 and dichromate—Cr_2O_7 which is more toxic to animals and plants than trivalent chromium.

- **hidden erosion**
slow mechanical erosion of loose soil.

- **hidden solonchak**
contains salts precipitated only within the soil; occurs in the case of deep water table.

- **high acre value crops**
are crops such as fruits, vegetables and tobacco with which the acre value of the produce is quite high.

- **high moisture grain**
any of the grains with a moisture content higher than that which would ensure safe storage. Usually somewhere around 13 per cent, wet-basis.

- **high pressure gate frame**
it is a hollow casting of iron or steel of desired cross-section having appreciable longitudinal length to form water passage upon upstream and downstream sides of the gate leaf.

- **high pressure gates**
high pressure gates are used in outlet works of dams. They are generally installed in tandem, with the upstream gate serving as an emergency gate and the downstream gate as a regulating gate. They are, often used as emergency gates in combination with needle valves and hollow jet valves. The gates may be operated as emergency gates under heads up to 250 feet (76.2 metres) and as regulating gates under heads up to 180 feet (54-86 metres). The gate is fitted with air inlets.

- **high salt water leaching**
sodium-affected soils do not disperse unless the electrolyte concentration is low. Saline-sodic soils usually have adequate infiltration until excess electrolytes are leached. The infiltration of water into alkali soils can be improved by increasing the electrolyte concentration of the soil solution because high water functions as a flocculent and a source of divalent cations for replacing sodium from the exchange complex of sodic soils.

- **high technology**
the use of ultramodern and electronics tools for performing tasks and controlling machinery and processes.

high water
the highest water level reached during a flood or a particular reservoir operation.

high water controls
controls that consist of the whole channel and flood plain over a considerable reach, or a large structure, such as a bridge, whose general configuration and dimensions are not seriously affected by any except the most extraordinary floods.

high water line
the line of the shore of a river, lake, or sea which is ordinarily reached at high water.

high water mark
a mark on a structure or natural object, indicating the maximum stage of tide or flood.

high water training or training for discharge
a type of training aiming at the provision of efficient cross-sectional area for the expeditious passage of the maximum flood.

high yielding variety (hyv)
are varieties which respond to higher levels of modern inputs like fertilisers and give higher yields.

highest stage of the water table
at a given place and for a given cycle of fluctuation, it is the highest altitude reached by the water table at the place in that cycle.

highway erosion control
prevention and control of erosion along the highways by mechanical and vegetative soil conservation practices.

hill planter
a planter which hilldrops seed.

hillock
a mound or a small hill.

histic epipedon
a surface horizon very high in organic matter and saturated with water at some time during the year unless artificially drained.

histograph
a map or chart of a river or drainage basin on which a series of time lines give the time of transit of water, originating on a time line, to flow down to the outlet of the system.

hitch angle
the angle between the line of pull and the line of draft.

- **hitch point**
  the point at which one piece of equipment is connected to another piece of equipment.
- **hive**
  place or box in which bees are kept.
- **hoe**
  implement for hoeing and interculture of crops.
- **hoe out**
  to remove weeds from a field with a hoe.
- **hoe type trenching machine**
  a machine with a stiff-arm digging bucket that loads toward the machine for making deep excavations. Commonly called back hoe.
- **hoes**
  are implements used for the hoeing and interculture of cultivated crops.
- **hog hook**
  a tool which consists of a short, iron shank, a ring or hand grip on one end, and a sharpened hook on the other. It is used in hog slaughtering and dressing to lift, move and hold the carcass.
- **hog house**
  any structure used for farrowing or sheltering of swine in any phase of the enterprise.
- **hog trough**
  a rack into which a slaughtered hog is placed to hold the carcass on its back to facilitate skinning.
- **hog, hogger**
  a mill or shredding machine.
- **hogback**
  a ridge formed by the outcropping edge of tilted strata; hence any ridge with a sharp summit and steeply sloping sides, as an esker.
- **hogringing crate**
  a wooden box-like device which is used to facilitate the job of nose ringing pigs to keep them from rooting.
- **hoist**
  the mechanism by which a bucket or blade is lifted, or the process of lifting it.
- **hoist line**
  the line which operates the hoist block and lifts the load. Also the line which raises or lowers the scaper body.
- **hoisting chamber**
  a chamber built at the top of a dam to house remote control hoisting mechanism.

- **hoisting plug**
connecting link between the hoisting line and the drill pipe, also used with casing plug or adapter to run or pull casing.

- **holding**
agricultural land that is being cultivated by an owner, tenant, or manager as one unit. Sometimes a number of farms are worked as one holding.

- **holding alley**
a confined area in a milking parlour system where cows are held prior to being milked.

- **holding area**
a section of a barn, shed or open lot where cows are confined prior to being milked.

- **holding ground**
the condition of the bottom of an anchorage area; called good or bad according to whether or not the material of which the bottom is composed will prevent a ship's anchor from dragging.

- **holding method**
a method of pasteurisation of milk in which the milk is preheated to about 143°F. and then pumped to a holding tank where it is held for 30 min. The milk is then cooled to about 40°F.

- **hollow clay tile**
a masonry unit having a cross-section of a square or rectangular tube, made of clay, cured and burned.

- **hollow dam**
a modification of buttress dam in which the buttresses are grouped or joined together leaving hollow spaces in between.

- **hollow jet**
a jet having an annular cross-section, with little or no flying spray, and the central core, composed of air, is ventilated.

- **hollow jet valve**
this valve is essentially a needle valve with the needle, or closure member, pointing upstream. The nozzle is eliminated allowing the water to discharge from the bell-shaped body in a tubular or hollow jet, the outside diameter of which does not change regardless of valve opening. This valve is, primarily, used in outlet works and is located on the downstream end of the outlet pipes. The water discharged may be closely regulated over the entire opening range of the above.

- **holocalcaire**
  soil in which the calcium carbonate comes from the soil itself.

- **holophyte**
  an organism that obtains food by photosynthesis.

- **holozoic organisms**
  organisms that prey on others, whether plant or animal, and thereby affect their population.

- **home grinder**
  a garbage grinder.

- **homeothermic mechanism**
  the means of balancing heat losses of animals or humans with heat gains.

- **homogeneous earth dam**
  an earth dam composed of a single type of material, except for protective material on the exposed faces.

- **homogeneous mass**
  a mass that exhibits essentially the same physical properties at every point throughout the mass.

- **homogenisation, efficiency of**
  this is determined by examining a sample of milk with a high power microscope and noting the diameter to 2 microns of the fat globules in comparison.

- **homogenised milk**
  milk which has been forced under pressure through a small orifice. The treatment breaks up the large fat globules into particles of small, uniform size so that they will not rise to the surface.

- **honeycombing**
  internal cracks in timber due to drying stresses.

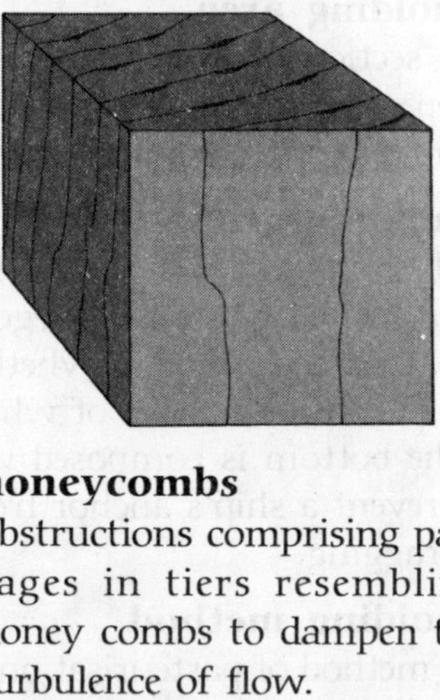

- **honeycombs**
  obstructions comprising passages in tiers resembling honey combs to dampen the turbulence of flow.

- **honeycomb structure**
  an arrangement of soil particles having a comparatively loose, stable structure resembling a honeycomb.

hood
it is the U-shaped body or structure made of reinforced concrete and used over the body wall to form a siphon duct.

hoof shovel
a cultivator shovel which has a triangular point.

hook
a spit whose outward end has been shaped by currents into a curve or hook; a cape or headland that turns inland at its extremity.

hook and point gauge
an adjustable gauge consisting of a pointed metallic rod which can be set with its tip at the liquid surface. If the rod points upward, the instrument is called hook gauge, while if the rod points downward, it is called a point gauge.

hook gauge
a pointed U-shaped hook attached to a graduated staff or vernier scale, used in the accurate measurement of the elevation of a water surface. The hook is submerged, and then raised, usually by means of a screw, until the point just makes a pimple on the water surface.

hook up
a typical arrangement of equipment for construction, development or testing of a well.

hooked groyne
an additional groyne built between existing groynes when it is found that they are spaced too far apart for mean water training.

hookworm
a nematode with hooks in its mouth, which is a parasite in man.

hopper
the box-like device on many farm machines in which materials are placed for processing or distribution or for temporary accumulation. For example, the box-like part of a hay baler in which the hay is placed before it is pressed into a bale; the box-like apparatus on the top of a grain drill which holds the seed; the box-like holding device on a feed grinder into which feed materials are placed for metering into a grinder; gathering or holding, funnelling device to direct material into materials handler.

- **hopper bottom bin**
a grain bin with a funnel shaped bottom to facilitate removal of the grain.

- **hopper bottom plate**
the solid plate on which the seed plate rotates in row crop planter.

- **hopper funnel**
a funnel that directs the silage into the feed rolls of a silage harvester.

- **hopper type clarifier**
a sample sludge blanket clarifier with the inlet at the bottom of the sludge hopper. The coagulated water flows up and through the sludge blanket. The surface loading is 40 to 100 m^3/day per m^2 of surface area of tank, depending on the quality of the raw water.

- **horizon A**
the upper horizon of the soil mass from which material has been removed by percolating waters; the horizon of eluviation; the surface soil, usually of dark colour due to the presence of organic matter.

- **horizon B**
the horizon of deposition, to which materials have been added by percolating waters; the horizon of illuviation; the subsoil.

- **horizon C**
the horizon of more or less weathered parent material underlying horizon B; the substratum.

- **horizons**
a series of soil layers overlying one another. Classified into three main categories: A horizon—topsoil; B horizon-subsoil, horizon of accumulation; C horizon—parent material; D horizon—underlying strata.

- **horizontal bench terrace**
a type of bench terrace which has no measurable slope from back to front of the bench.

- **horizontal centrifugal pump**
a centrifugal pump in which the impeller shaft is in a horizontal position and usually necessitates priming to start it.

- **horizontal control**
basic framework of points whose horizontal position azimuth and interrelationship have been determined accurately.

▪ **horizontal creep**
the creep in horizontal direction.

▪ **horizontal grill type sand trap**
a type of sand trap in which the water is let into a sediment basin immediately above the weir. The water then passes through horizontal grill fixed into the sediment basin and drops its coarse sediment into it, which is sluiced away. De-sedimented water passes over the weir.

▪ **horizontal revolution in agriculture**
increased land use by expanding cultivated land area through the utilisation of fallow and marginal lands and reclaiming culturable waste lands, thereby increasing land use intensity.

▪ **horizontal well**
a tubular well pushed approximately horizontally into a water-bearing stratum, or under the bed of a lake or stream.

▪ **hormone herbicides**
chemicals which exert an exterminatory effect upon the growth processes in such a way as to cause death of plants.

▪ **horsehead**
a light framework (or head frame) used over a pit for supporting a pulley block for use in hoisting and lowering men and materials.

▪ **horsehoe bend**
a river bend such that only a neck of land is left between two parts of the stream. The river may cut through, leaving a crescent-shaped lake.

▪ **horsepower**
1. the unit of measure for power in the US system of measurement.
2. the amount of power required to do 33,000 foot-pounds of work in one minute.

▪ **horsepower rating (HP)**
the rated horsepower output for an electric motor or internal combustion engine.

▪ **hose**
a strong flexible pipe made of leather, canvas, rubber, etc., and used for the conveyance of water, or air under pressure, to any particular point.

▪ **hour meter**
panel instrument that records the hours of operation for

many agricultural machines, especially tractors.

- **housing pipe**
a larger diameter plain pipe housing the borehole pump bowls in case of compound wells.

- **hovers**
a sheltering cover for chicks, pigs or lambs, usually placed inside a regular building housing them. It may be heated artificially when needed.

- **Howard method for worm count**
a quick test for fragments of insects in vegetable products.

- **howell hunger valve**
a valve having two telescopic cylinders with a streamline dispersing cone secured to the inner cylinder by radial webs. The apex and axis of this cone are coincidental with the centre line of the conduit and valve body, and its apex faces upstream. The outer cylinder closes the sideway opening between the cone and the inner cylinder when it is slided in position. In its open position, the water is discharged on the sides of the cylinder in the form of highly diverging nappe hollow inside, in the shape of the surface of a cone.

- **hub crop**
hub crop is a sequential cropping system which has the greatest comparative advantages over other crops, e.g., vegetables/truck crop cultivated near city markets.

- **hub spline**
that part of a power take-off unit which fits onto the shaft. It has grooves on the inside diameter which match the spline-ways on the shaft. The power is transferred normally from the shaft to the hub.

- **huller**
a machine used for hulling. While removing the hull, the machine removes part of the bran also.

- **huller picker**
a cotton harvester which picks and removes the lint from the seed in one operation.

- **huller rib**
a device on an air-blast cotton gin which holds back the hulls, but allows cotton to pass into the roll box.

- **hulling**
it is the process of removing hull or husk and bran in one operation.

- **humid**

an adjective applied to regions or climates where precipitation is adequate in amount and occurs at such times that agriculture can be carried on without irrigation, but where drainage may be required to remove excess moisture; a climate with sufficient precipitation to support a forest vegetation.

- **humidity**

a general term indicating the amount of water vapour in the air.

- **hump stilling basin**

when the centre line of a jet emerging from an outlet is below the stream bed elevation, but not too far to be completely submerged by the tail water, a hump in the stilling basin floor may be provided to spread the jet and permit the formation of a stable hydraulic jump. The stilling basin in this case is called hump stilling basin, or hump basin.

- **humus**

the dark or black carboniferous residue in the soil resulting from the partial decomposition of plants and animate originally growing therein. It is the organic portion of soil. It is highly colloidal, has high cation exchange capacity, will absorb much water with swelling, shrinks on drying, but does not have the pronounced cohesion of clay mineral colloids.

- **humus sludge**

biological film that has sloughed off a trickling filter and settled out in the humus tank.

- **humus tank**

a secondary sedimentation tank that purifies the effluent from trickling filters by separating the humus sludge from it.

- **hurdle**

a sediment arresting device. The hurdle consists essentially of a row or rows of pillings; the piles driven either singly or in clumps, the piling being connected lengthwise of the hurdle by wattling of fine brush or by curtains composed of brush and lodged against the upstream side of one of the rows of piling, the whole forming a permeable dyke through which silt-laden current can pass, though with greatly diminished velocity, resulting in deposit of silt above and below the hurdle.

- **hurdling**
  the action of water flowing over the top of a baffle or baffles with a high velocity, like a horse taking a fence.
- **hydration**
  the chemical process of a material combining with water to form a hydrate.
- **hydraulic action**
  the mechanical loosening and removal of materials by water alone.
- **hydraulic actuator**
  a hydraulic device for converting fluid energy into mechanical energy.
- **hydraulic agitator**
  a device which keeps the tank mix from settling out by means of water flow under pressure.
- **hydraulic cylinder**
  a type of hydraulic actuator that produces linear motion.
- **hydraulic equilibrium of water in soil**
  the condition for zero flow rate of liquid or film water in the soil. This condition is satisfied when the pressure gradient force is just equal and opposite to the gravity force.
- **hydraulic fill**
  an adjective applied to an earth structure or grading operation in which the fill material is transported and deposited in place by means of water pumped through a pipe line.
- **hydraulic fill dam**
  dam built from earth material pumped into place with water, the fines being washed towards the centre for ensuring more imperviousness.
- **hydraulic grade**
  in a closed conduit under pressure or an artesian aquifer, or ground-water basin, a line joining the elevations to which water would rise in pipes freely vented and under atmospheric pressure. In an open channel, the hydraulic grade is the water surface.
- **hydraulic mechanism**
  the equipment or mechanism used in the transmission and control of forces through hydraulic systems. Basically composed of a pump, valves, motor, cylinder, reservoir and fluid or oil.
- **hydraulic motor**
  a type of hydraulic actuator that produces rotary motion.

- **hydraulic pump**
a mechanical pump used to transmit and control hydraulic fluids under pressure in a closed circuit.

- **hydraulic ram**
a device or machine for raising water by utilising the momentum of water flowing by gravity intermittently through a pipe to lift a portion of the water to an elevation greater than that of the source of supply.

- **hydraulic shutter**
an inclined shutter supported by a plunger, which has a roller at the top. The plunger is worked by a hydraulic ram. The roller slides along the shutter and operates the shutter.

- **hydraulics**
the branch of physics that deals with the mechanical properties and applications of liquids in motion.

- **hydro cooling**
the process of removal of heat by immersing in cold water.

- **hydrocarbon fuel**
engine fuels consisting of hydrogen and carbon.

- **hydrodynamic lubrication**
a method of engine lubrication where movement of the lubricant is by adhesion to moving or sliding surfaces.

- **hydrodynamics**
use of liquids at high flow and low pressure to perform work.

- **hydroelectric**
a method of generating electricity using the energy of moving water to turn the generator turbine.

- **hydroponics**
the practice of growing plants without soil.

- **hydrostatic lubrication**
a process of engine lubrication where movement of the lubricant is due to a pump.

- **hydrostatics**
use of liquids at low flow and high pressure for performing work.

- **hygroscopic water**
water that is held too tightly by soil particles for plant roots to absorb.

- **i.n.k. a. process**
a method of Swedish origin for operating aeration tanks in the activated sludge process.

- **ICOLD**
the International Commission on Large Dams, a body that is particularly interested in dam safety.

■ ideal irrigation interval
the length of time the net amount applied by irrigation will last before the crop uses this moisture under peak use conditions and another irrigation is required.

■ ideal soil
a soil which conforms to the idea one has of the normal evolution through the action of the factors present under natural conditions.

■ idle adjusting needle
a screw with needle point which meters the airfuel ratio of a carburettor while the engine is operating at slow or idling speed.

■ idle circuit
in the carburettor, the passage through which fuel is fed when the engine is idling.

■ idle land
land which has been cultivated, but put to disuse; abandoned land; fallow land.

■ idle mixture
the air-fuel mixture supplied to the engine during idle.

■ idle mixture adjustment screw
the adjustment screw that can be turned in or out to lean or enrich the idle mixture.

■ idle port
the opening into the air horn through which the idle circuit in the carburettor discharges.

■ idler pulley
a pulley or mechanism to tighten a belt or chain running between driver and driven pulleys, gears or other mechanisms. Idler pulley is located on the non-driver side of the belt or chain.

■ idling
the slow rate of movement of an automobile.

■ idling speed
the speed at which the engine runs without load when the accelerator pedal is released.

■ idling speed adjustment
a stop-screw adjustment on the throttle butterfly shaft of a carburetor to regulate the idling speed of an engine.

■ ignition system
engine auxiliary system for initiating the combustion of the air and fuel mixture.

■ ignition timing
the point in the engine cycle when combustion of the air and fuel mixture occurs.

■ ihp
indicated horsepower. The horsepower developed within

the cylinders of an engine as ascertained from the displacement pressure record on an indicator diagram along with engine rpm.

- **illite**
a group of clay minerals similar to montomorillonite but containing potassium, consisting of aggregates of small flakes, and possessing less swelling and expansive properties.

- **illuvial horizon**
horizon in a soil profile in which soil material carried from an upper horizon has been deposited as a result of precipitation from solution or separation from suspension.

- **illuviation**
the process of deposition of soil material removed from one horizon to another horizon of the soil usually from an upper horizon to a lower horizon in the profile. It is a soil development process that involves translocation of humus, chemical substances, fine mineral particles, etc.

- **illuviation cutans**
cutans formed by movement of the cutanic material in solution or suspension and subsequent deposition. A sharp boundary between the cutanic and non-cutanic material is indicative of an illuviation cutan.

- **imbibational water value**
the difference between the moisture equivalent and the xylene equivalent of a soil, which is a measure of the soil's ability to imbibe water.

- **imbibitional moisture**
moisture over and above needed to satisfy the hygroscopicity of soil particles. Turns soil visibly moist by the colloidal clay imbibing moisture; clay particles increase in size and swell. Held fast and tenaciously by clay, but available to plants for absorption.

- **Imhoff cone**
a conically shaped graduated glass vessel used to measure approximately the volume of settleable solids in various liquids of sewage origin.

- **Imhoff tank**
a deep two-storyed tank originally patented by Karl Imhoff, consisting of an upper or continuous flow sedimentation chamber and a lower or sludge-digestion chamber. The floor of the upper chamber slopes steeply to trapped slots, through which solids may

slide into the lower chamber. The lower chamber receives no fresh sewage directly but is provided with gas vents and with means for drawing digested sludge from near the bottom.

- **immature plants**
those that are green or have not yet matured. Suitable for mixing with soils as green manure.

- **immature profile**
a soil profile that is still developing.

- **immature soil**
a soil with indistinct or only slightly developed horizons because of the relatively short time it has been subjected to the various soil forming processes. A soil that has not reached equilibrium with its environment.

- **immiscible**
a term used widely in the fertiliser trade for mixtures and liquid fertilisers. Substances of same phase that cannot be mixed or blended are termed as immiscible. In case of liquids immiscibility is due to difference in chemical nature. Solids, which differ much in their physical properties, could also be immiscible.

- **impact**
the striking together of two or more masses when particles or streams of water suffer impact, energy loss, or more accurately, energy transformation results.

- **impact stilling basin**
a type of stilling basin initiated by the need for relatively smaller basins to provide energy dissipation independent of tail water variations or any tail water at all.

- **impedance**
the total opposition to current flow in a circuit.

- **impeded drainage**
a condition which reduces the movement of water through the soil under the influence of gravity.
2. impeded removal of excess water from all or some portion of the soil profile producing some limitation of plant growth or tillage of the land.

3. loose material which can readily be scoured or deposited; the material may range from very fine sand and silt in rivers and canals with small slopes to gravel, pebbles and boulders met in torrents.

- **impeller**
  1. the rotating element of a pump which imparts kinetic energy to the fluid and produces desired head.
  2. a rotating set of vanes in a pump, etc., designed to impel rotation of a mass of fluid; on the peripheral speed of the vane tips depends the head produced and the working pressure of a pump.
- **impeller blower**
  a blower or fan giving centrifugal action to materials in addition to a wind blast for the movement or conveying of materials, i.e., a silage blower moving chopped corn stalks into a silo.
- **impeller pump**
  any pump wherein the water is moved through the continuous application of power by some mechanical agency or medium.
- **impeller, rotor**
  the rotating wheel carrying blades that move the water or air in a centrifugal pump or axial-flow pump or fan.
- **imperfect humus**
  humus in the initial stages of formation. It is not well mixed with soil but resting on it.
- **impermeability**
  the property of rock, earth or other material which prevents perceptible movement of water through it under pressures ordinarily encountered in subsurface water. Even though the material allows capillary movement, it still may be impermeable.
- **impermeable**
  preventing passage as in the use of an impermeable membrane.
- **impermeable confining bed**
  compare with permeable confining bed. There is probably no confining bed that is strictly impermeable over any wide area.
- **impermeable fixed weir**
  a fixed weir whose water bearing surfaces are of watertight construction. It may be built with timber, masonry, concrete (plain or reinforced), steels or combinations thereof.

impermeable rock
a rock whose texture is such that water cannot move perceptibly through it under pressures ordinarily found in subsurface water. While the rock may be impermeable under such ordinary pressure, it may allow movement at higher, pressures, or because of capillarity.

impermeable weir
a weir made up of impermeable material, such as rock stone, brick masonry, concrete, steel, timber.

impervious
highly resistant to penetration by water and usually resistant to penetration by air and plant roots. Impenetrable. In field practice the term is applied to strata or horizons that are very slowly penetrated by water and that retard or restrict root penetration.

impervious bed
a bed or stratum through which water will not move under ordinary hydrostatic pressure.

impervious soil
a soil through which air, water or plant roots penetrate very slowly if at all.

imperviousness coefficient
the ratio, expressed decimally, of effectively impervious surface to the catchment area under consideration.

implement shed
a farm building used primarily as a storage place for farm implements, frequently including an area where machines may be repaired.

implement tool
a machine or piece of equipment operated by extension from the basic machine, i.e., a plough or other tool operated by or from a tractor.

implements
the equipment used on farm to carry out different farm operations either with the help of tractor or bullock power.

implements for carrying soil
they are wheel barrows, keni or levelling karah, leveller and clod crusher and bund formers.

importation works
works constructed in connection with an irrigation system when water is secured from another drainage basin. They generally in-

clude a diversion structure or collection canals at the source of the supply, and a tunnel through the divide, or link canals.

- **impound**
  1. to collect water as by damming a stream for irrigation purposes, or the like.
  2. a reservoir for water, as one made by damming a stream.

- **impounding dam**
  dam used to withhold water giving stability or withholding water for other purposes.

- **impounding reservoir**
  a reservoir with gate controlled outlets wherein surface water may be retained for a considerable period of time and released for use at a time when the ordinary flow of the stream is insufficient to satisfy requirements.

- **impoundment**
  artificial storage of water as a reservoir, dug out or sump etc.

- **impoverished soil**
  soil which has been continuously cropped and has had little or no care so that it is incapable of supporting a satisfactory or economic crop.

- **improved land**
  land that has had its value increased either because it has been reclaimed from waste for agriculture or because it has had its yield potential increased by such methods as drainage or levelling.

- **improved production technology**
  the use of improved varieties and associated improved crop management practices followed on a farm.

- **improved water wheel**
  a modification of the paddle wheel in which the lower parts of the paddles are placed in a close fitting box to reduce spilling of the water from the edges of the paddle, and making it possible to lift water to comparatively greater heights. The labourers sit on a wooden or bamboo scaffolding erected over the installed wheel whose paddles they push down with their foot; known in India as chakram.

- **improvement cutting**
cutting of undesirable species, unsatisfactory growth of plants having insect infestation or disease for the purpose of improving the quality and vigour of other trees in an immature stand.

- **impulse pump**
a type of pump which raises water by periodic application of a force suddenly applied and suddenly discontinued. The hydraulic ram is the most common example of such type.

- **impulse reaction turbine**
a turbine utilising both the impulse and reaction principles.

- **impulse turbine**
a type of turbine in which all the available energy in the fluid is converted into kinetic energy before it acts on the runner of the turbine.

- **in and out method**
a method of bank strengthening, generally followed in low ground, in which subsidiary (additional) banks are constructed outside the original canal banks and parallel to them, with cross banks at suitable intervals.

- **in season**
that part of the year when a production is normally harvested and is thus cheaper and more plentiful.

- **in situ**
in the natural or original position of location.

- **in situ water harvesting**
a technique of water harvesting where water is stored in the soil profile itself.

- **in the row applications**
a method of fertiliser application wherein fertiliser may be run down the same tube with the seed or it may be placed in the furrow with the seed. It also includes application of fertiliser at the bottom of deep furrow followed by covering the before planting.

- **inactive storage (capacity)**
storage capacity, exclusive of dead storage, below which evacuation is not contemplated because of minimum irrigation and power load requirements or of operating agreements not to draw the reservoir below a given capacity or elevation.

- **incandescent light**
a type of lamp that glows because of the heat produced as

current flows through a high resistance tungsten filament.

- **inceptisols**
  soils that are generally moist, having pedogenic horizons of alternation of parent material but not of illuviation.

- **incidental recharging**
  natural replenishment of ground water from the seepage of conveyance systems, deep percolation from normal irrigation, sewage, cooling water and other wastes.

- **incipient erosion**
  the early stages of erosion especially with reference to gullying.

- **incipient wilting**
  small loss in turgidity during warm weather, even when soil is moist. It may not lead to any visible stress symptoms like rolling or folding of leaves and is usually reversible at night.

- **incised river**
  a river which has cut its channel through the bed of the valley floor, as opposed to one flowing on a flood plain.

- **inclined column grain dryer**
  a grain dryer in which the passage or column holding the grain for drying is in an inclined position.

- **inclined plate seed metering device**
  the seed plate of a metering device positioned at an inclined angle or axis to a horizontal position.

- **inclined tube settling tank**
  a type of sedimentation tank in which water or sewage flows upwards through settling tubes.

- **inclined tunnel spillway**
  a spillway comprising a steeply inclined tunnel connected to the reservoir and having control devices (gates) at its inlet end.

- **included water**
  water which is in the clods of the soil.

- **income equivalent ratio (ier)**
  the ratio of the area needed under sole cropping to produce the same gross income as one hectare of inter-cropping at the same management level. IER is the conversion of LER into economic terms.

- **incomplete fertiliser**
  a commercial fertiliser not having, all three of the fertiliser elements.

- **incomplete soil**
  soil which has not arrived at maturity, which does not have the characteristics horizons which it presumably ought to have.
- **incorporate into soil**
  the mixing of a pesticide into the soil by mechanical means.
- **incorporating or mixing**
  tillage operations which mix or disperse foreign materials, such as chemicals or plant residues into the soil.
- **incorporation by reference**
  including, as an official part of the contract drawings, references, specifications or other data by referring to known building codes, departmental manuals, standard specifications, standard designs and even previously executed contracts.
- **increment in land revenue**
  a charge added to land rent payable to the government by cultivators and is based on increased benefit derived annually by the cultivators on account of provision of irrigation supplies, it sometimes takes initially the form of 'water advantage rate', or 'canal advantage rate' and is merged with the land revenue at the settlement following the construction of a canal.
- **incremental loading**
  step aeration of an aeration tank.
- **incrustation**
  any clogging or stoppage of a well screen, including the water bearing formation surrounding it, which is the result of a deposition or collection of material in, on and around the openings of the screen and the voids in the water bearing formation in which the screen is placed.
- **incubation**
  preliminary wetting of organic materials added to the soil in artificial recharging for increasing and maintaining recharge rates for their decomposition, after which they exhibit their beneficial results.
- **independent power take off**
  a power distribution system used on tractors in which the transmission and power take off are operated through independent clutches. Either the transmission or power take off are operated without affecting the other.

- **independent trials**
  the successive trials of an event are said to be independent if the probability of outcome of every trial is independent of the outcome of the others.

- **indeterminate growth**
  the flowers are borne on lateral branches, the central stem continuing vegetative growth with blooming continued for a long period like in cotton.

- **index**
  the average rainfall intensity above which the volume of rainfall equals the volume of observed runoff and equals total basin discharge divided by the duration of rainfall, provided that the rainfall intensity is continuously in excess of the index.

- **index map**
  map of a tract, area or country showing contours, location of roads, rivers, canals, cities, towns, villages and other topographical features, generally to a small scale to cover a large area.

- **index of association**
  an indicator of the proper measurement of association of the properties of statistical data, e.g., the index of association of column with row of a table.

- **index of clay transport**
  the ratio of the clay content of the accumulation horizon B to that of the leached A horizon.

- **index of wetness**
  the precipitation/rainfall/runoff for a given year expressed as ratio to the mean annual precipitation/ rainfall/runoff over a considerable period of time.

- **index plate**
  a plate perforated with holes, arranged in concentric circles, used in a milling machine.

- **indicated power (Pj)**
  the theoretical power output of an engine based on mean cylinder pressure, engine displacement, and engine operating speed.

- **indicator**
  an instrument by which the working steam records its

working pressure from which the power of the engine may be calculated.

- **indicator diagram**
a diagram which represents the pressure on an engine piston at the various piston positions throughout the stroke.

- **indicator plant**
a native plant that indicates, in general, and often in a specific manner, the nature of soil conditions with regard to moisture and salinity. Dominant species are the most important indicators of such conditions.

- **indicator plot**
a small plot kept under detailed observation within a large experimental area, e.g., when testing regeneration methods.

- **indicator post**
a post attached to the operating mechanism of a valve or gate, and designed to show whether the valve or gate is open or shut.

- **indirect benefits**
the values added over and above the value of the immediate products or services of the project as the result of activities stemming from or induced by the project.

- **indirect damage**
losses resulting from floods but not from direct action of flood water; for example, losses resulting from the interruption of the production of goods and services.

- **indirect discharge measurement**
a determination of the peak discharge, other than by current meter, by the slope area, contracted opening, culvert, based on a survey of high water marks following a flood.

- **indirect flood damage**
flood damages other than direct flood damage to property, structures, crops, etc., such as loss of business, costs of emergency measures, etc., to monetary evaluation.

- **indirect pumping**
pumping of water into elevated storage reservoirs, whence it flows by gravity into the distribution system.

- **individual beam shear pins**
when an obstruction is encountered with the plough bottom, the bottom shears the soft upper holding pin and pivots to the rear, clearing the obstruction on the lower bolt.

The sheared pin or bolt must be replaced to maintain proper operations.

- **individual branch circuit**
a branch circuit, which serves a single electrical load.

- **individual equipment circuit**
a branch circuit installed to supply a single motor or other appliance.

- **individual farrowing house**
a separate building equipped so that sows that are to farrow may be kept isolated in clean and quiet individual pens at farrowing time.

- **individual project**
single or multi-purpose project, without consideration of the integrated river basin development.

- **Indore process**
composting in hot climates (originally Indore, India) using layers of night soil and vegetable waste alternately. It is kept aerobic by regularly turning for two or three months and then is left to mature.

- **induced benefits**
profits of all enterprises from supplying goods and services for the increase in farm purchases for family living and production expenses.

- **induced crusts**
compacted soil at the soil surface about 1 cm thick caused by raindrop splash.

- **induced draught fan**
a fan that sucks flue gases out of a furnace and pushes them up the chimney. It is therefore placed between the boiler and the chimney. Since hot flue gases pass through it, it is liable to be corroded and its service conditions generally are more severe than of a forced-draught fan. Fans made of stainless steel can work at temperatures up to 600°C, provided that the bearings are water-cooled.

- **induced recharge**
additional increment to an aquifer from surface waters and precipitation obtained by providing increased storage capacity by abstraction of water by pumping or other means. It is distinguished from 'artificial recharge' in that it does not involve the conveyance of water to some part of area where it enters the ground.

- **induced recharge of an aquifer**
discharge of water from a stream into an aquifer.

- **inductance (L)**
the property of an electrical circuit that opposes any change in current flow.

- **induction**
action of producing a voltage in a conductor or coil by moving the conductor or coil through a magnetic field (or moving field past conductor or coil).

- **induction motor**
a kind of AC electric motor where current flow in the rotating motor component is caused by induction.

- **inductive reactance**
the amount of opposition to change in current flow produced by an inductor.

- **inductor**
a coil placed in an electrical circuit to produce a desired amount of inductance.

- **indurated**
rocks that have been hardened by heat. Occasionally the term is applied also to sediments hardened by impregnating solutions.

- **indurated soil**
very strongly cemented; brittle, does not soften under prolonged wetting, and is so extremely hard that for breakage a sharp blow with a hammer is required; hammer generally rings as a result of the blow.

- **induration**
a geologic term describing the process of hardening undergone by sedimentary materials, largely due to cementation by mineral matter deposited from solution in water. The resulting material is usually very hard and rigid.

- **industrial river**
a river whose flow in dry weather consists mainly of effluents from domestic or industrial wastewater treatment plants.

- **ineffective rainfall**
that portion of the total annual or seasonal rainfall which is lost by surface runoff, deep percolation beyond root zone, the moisture remaining in the soil after the harvest of the crops and is not useful for the next crop. The period considered here is from the start of the first tillage operation till the harvest and not

only the actual crop growing period.

- **inert**

a compound or element that does not readily react with other elements or compounds.

- **inert matter**

in seed testing, it refers to all foreign matter such as places of broken or damaged seeds, empty glumes, lemmas, paleas, soil, sand, chaff, stems etc., and all other matter not seeds.

- **infield**

that part of a farm, usually in a hilly district, that is close to the farmhouse. The outlying area, or out-field, is likely to consist of rough grazing or moorland with occasional arable patches. The infield is normally fenced, manured, and cultivated.

- **infiltrability of soil**

the flux of water the soil can absorb through its surface, in contact with free water.

- **infiltrate**

1. water containing salts obtained after it has passed downward through a body of soil.
2. the penetration by a liquid or gas of the pores and interstices.

- **infiltration**

1. the flow or movement of water through the surface into the soil body or ground.
2. the absorption of liquid water by the soil, either when it falls as rain, or when applied as irrigation, or from a stream flowing over the ground.
3. flow from a porous medium into a channel, pipe, drain, reservoir or conduit.

- **infiltration capacity**

the maximum rate at which soil can absorb rainfall in a given condition. Infiltration decreases as the rain continues, because some of the soil particles swell and block the pores; also some of the pores fill with capillary water. Infiltration tends to increase with porosity.

- **infiltration capacity curve**

curve showing what the infiltration rate would be at any period during a specific storm if the rainfall intensity were to equal or exceed the capacity at that instant.

- **infiltration coefficient**
the ratio of infiltration to rainfall.

- **infiltration complex**
a group of areas which have comparable similar physical conditions of soil, slope, litter, and cover, and which have a similar infiltration rate.

- **infiltration ditch**
an artificial ditch which extends into the zone of saturation and through which water flows by gravity from the zone of saturation to the land surface or into a well.

- **infiltration diversion**
diversion of water from a stream by means of perforated pipes or other previous conduits laid under the stream bed.

- **infiltration gallery**
a tunnel or well dug near a river to withdraw water from it. The water is partly filtered as it flows through the soil between the river and the gallery.

- **infiltration head**
the head of water which makes it flow or infiltrate through the media.

- **infiltration index**
the average rainfall intensity over and above which the mass of rainfall (the supra-rain) equals the mass runoff.

- **infiltration rate**
1. the rate at which water penetrates the surface of a soil. The term is usually used to refer to water occurring as precipitation, but is also applied to water flowing or standing upon soil.
2. the rate at which ground water enters an infiltration ditch, gallery, drain, sewer or other underground conduit.

- **infiltration rate curve, or recharge rate curve**
1. curve showing recharge rate variation with time.
2. curve showing the actual rate of infiltration during a particular storm.

- **infiltration tunnel**
an artificial tunnel having same properties as an infiltration ditch.

- **infiltration velocity**
the actual rate at which water is entering the soil at any given time. It may be less than the maximum infiltration rate because of a limited supply of water.

- **infiltration volume**
rainfall, expressed in millimetres (inches) depth

over a given area, that passes below the root zone of plants and ultimately reaches the zone of saturation.

- **infiltration water**
  water that has leached from the ground into a structure.

- **infiltro meter**
  a device by which the rate and amount of water percolating the soil is determined by measuring the difference between the amount of water applied and that which runs off. The water may completely cover the ground inside an infiltrometer ring or may be applied by a sprinkling mechanism (rain simulator) which provides a rather uniform sprinkling of water to a prescribed area at prescribed rate and size drops (impact) auxiliary apparatus consisting of a rain gage (either total or intensity) and a catchment basin or receptacles in which surface runoff is measured either as to rate or total flow. Infiltration or amount absorbed by soil is usually expressed in inches (of water) per standard interval of time.

- **inflammability limit**
  the lowest temperature at which a material will ignite and burn in favourable surrounding conditions.

- **inflatable weir**
  a weir constructed of neoprene reinforced with nylon and anchored to a concrete foundation. The weir inflates by water pressure and automatically deflates when the upstream water reaches a prescribed level.

- **inflation pressure**
  the amount of internal pressure above atmospheric within a container such as a balloon, tire, diving suit, or airplane fuselage.

- **inflow**
  the flow of water into a reservoir or structure.

- **inflow constriction**
  a constriction inserted at the inlet to a closed conduit.

- **inflow hydrograph**
  the graphical representation of inflow rate against time.

- **influence area**
  the surface area surrounding a well, or group of wells, during pumping, within which the water table or other piezometric surface is lowered by withdrawal of the water. The area varies in extent with the rate and duration of pumping.

- **influence value**
  the value of the portion of a mathematical expression that contains combinations of the independent variables arranged in dimensionless form.
- **influent**
  description of fluid that flows in; usually water, sewage, etc., entering a treatment system.
- **influent channel**
  channel delivering water to the settling, basin.
- **influent seepage**
  movement of gravity water in the zone of aeration from the ground surface towards the water table.
- **influent stream**
  a stream type, common in an arid climate, that loses water into the ground, thus recharging the groundwater, unlike the effluent streams that are usual in wetter countries.
- **influent water**
  subsurface water due to penetration into the lithosphere of surface waters, or atmospheric precipitation through large interstices or fractures.
- **influent weir**
  a weir at the inflow end of a sedimentation basin or other hydraulic structure.
- **influx**
  the gross uptake of ions by a plant root.
- **infra red analyser**
  an apparatus used to measure either CO_2 or water vapour content of an air sample. They depend on the differential absorption of radiation in 2 sample tubes with identical geometry so that it clearly records vertical profile of the above parameters in and above the crops or vegetation.
- **infra red thermometer**
  used to measure leaf and canopy temperature of plants. It collects the thermal energy emitted by the plant surface by fixed focus optics and converts into electrical signal by a sensitive detector, which is electrically conditioned for reading in temperature units. It is quite fast and collects energy at the speed of light.
- **inglis type fall**
  a flumed meter type fall with dimension standardised at the Central Water and Power Research Station, Poona (India). In this fall, use is made of horizontal impact for energy dissipation. The jump is held stable by means of a baffle.

- **ingot iron**
very low carbon steel generally made in the open hearth containing very small amounts of other elements.

- **ingrown meander**
a type of entrenched stream meander that 'grows in place' as the stream sinks its channel into bedrock. The meander pattern is not inherited from a previous erosion cycle and merely preserved as the stream cuts downward.

- **inherited profile characteristics**
are those inherited from the parent material and hence geological in nature, e.g., the presence of quartz, mica and feldspar in soils.

- **inhibitory toxicity**
in a bio-assay ASTM 31 defines this as any reduction by a pollutant of the rate of reproduction of diatoms, within 7 days.

- **initial abstraction**
the sum of interception and depression storage.

- **initial abstraction retention**
total amount of rainfall that may fall without causing a significant amount of direct runoff.

- **initial consolidation**
a comparatively sudden reduction in volume of a soil mass under an applied load due principally to expulsion and compression of gas in the soil voids preceding primary consolidation.

- **initial cultivation**
first cultivation generally by the use of heavy disc plough.

- **initial detention**
the part of precipitation which does not appear either as infiltration or as surface runoff during period of precipitation or immediately thereafter; includes interception by vegetation cover, depression storage and evaporation during precipitation; does not include surface detention.

- **initial flood**
flood caused by a continuous light rainfall preceding the major storm.

- **initial foliage storage**
the maximum amount of precipitation that is necessary to satisfy the foliage deficiency before some of it drips to the ground under the trees or another foliage plant.

- **initial intake rate**
  rate at which water will enter the soil when water is first applied; mm/hr.

- **initial loss**
  in hydrology, rainfall preceding the beginning of surface runoff. It includes interception, surface wetting, and infiltration unless otherwise specified.

- **initial outlay**
  expenditure incurred on a work or a system, at the time of construction excluding cost of subsequent extensions and improvements.

- **initial rain**
  the rain that falls in the beginning of a storm before the depression storage is completely filled is called the initial; that falling near the end at a rate less than infiltration capacity is called residual rain; the intervening period is the net supply interval.

- **initial set**
  when the paste (cement water) loses its plasticity and becomes sufficiently coherent to withstand a certain arbitrary pressure, it is said to have acquired initial set.

- **initial setting time (of cement)**
  the period elapsing between the time when water is first added to cement to form a paste and the time when that paste ceases to be fluid and plastic to a specified degree under the specified conditions of test.

- **initial soil storage**
  the maximum amount of precipitation that is necessary to satisfy the moisture deficiency under specific conditions of soil before any runoff takes place.

- **initial storage**
  volume/volumes of water in a reservoir/ multiple-reservoir system at the beginning of the particular period of analysis.

- **injection**
  placement of fertiliser below the soil surface by injection: usually refers to application of anhydrous ammonia in this way to avoid gaseous loss.

- **injection aerator**
  an aerator that injects air directly into a main, thus avoiding the loss of pressure that occurs with spray aerators or cascade aerators, when the

water is released from the pressure of the water main. A disadvantage of injection aeration is that excess air may bubble out at the consumer's tap.

- **injection type recharge projects**
  projects designed to inject underground, using pits, shafts, wells, etc., and involve the natural (as in stream beds) or artificial removal of soil to expose coarse or aquifer material; the corresponding methods are called injection methods (recharge).

- **injection well**
  a disposal well or a well for artificial recharge.

- **injection, compression ignition**
  the introduction of fuel by force, into an enclosed cylinder as in a diesel internal combustion engine. Injection starts as the piston nears the top of its stroke and heat of compression causes ignition of the fuel and burning continues throughout the duration of the injection.

- **injection, spark ignition**
  the introduction of fuel by force, into an enclosed cylinder such as in an internal combustion engine. In this case, injection is early in the stroke at low pressure. Nearing the extreme point of compression an electrical spark is produced causing a burning or explosion of the mixture of fuel and air.

- **injector**
  1. a nozzle and back-up valve used to inject fuel into an internal combustion engine cylinder.
  2. a hydraulic device designed to force water into a steam boiler under pressure.

- **inlet**
  1. a surface connection to a closed drain.
  2. a structure at the diversion end of a conduit.
  3. the upstream end of any structure through which water may flow.
  4 .a cross-drainage work consisting of an opening, in a canal bank suitably protected, to admit upland drainage water into the canal.
  5. a structure admitting water supplies from the source and leading them to the intake structure.

- **inlet and outlet**
  a type of drainage work in which the drainage water is

taken in and escaped out at a suitable site a little downstream along the canal. The outlet is essentially combined with some other work where an arrangement for escaping is, in any case, to be provided or may be added at small cost. The bed levels of the inlet/outlet and the canal need not be nearly the same.

- **inlet control**
a structure built at the upstream end of the outlet conduit of the outlet works and housing the regulating or emergency gates usually combined with the trash rack structure.

- **inlet structure**
a structure at the inlet of a spillway or intake of water supply system.

- **inlet time**
the time required for storm water to flow from the most distant point on a drainage area to the point where it enters a storm drain.

- **inlet well**
1. a well or opening at the surface of the ground constructed to receive surface water which is thence conducted to a drain.
2. a chamber which serves as a suction well for pumps in a pumping station.

- **inlet, or mouth**
1. the normal distance between the entrance lip of hood and the body wall.
2:the section between the entrance lip of hood and the body wall through which water enters the siphon.

- **in-phase**
condition existing in an AC electrical circuit when voltage and amperage reach peak positive and negative and zero values at the same time during the cycle.

- **input device**
computer hardware used to refer information into the computer. For example, a keyboard or mouse.

- **insulated-gate field-effect transistor (IGFET)**
a variety of field-effect transistor that operates at high speed with low power requirements.

- **insulation class**
a system that is used with electric motors for indicating the maximum safe operating temperature that will not cause premature damage to

the insulation on the internal windings.

- **insulator**
  a substance that has an extremely high resistance to the flow of electrical current.
- **intake**
  1. the process of bringing a new air or air-fuel mixture into the engine combustion chamber.
  2. the first stroke of the four-stroke engine cycle.
- **integrated circuit**
  a complete electronic circuit contained entirely on a single chip.
- **interactive process**
  a non-linear process where decisions made at one point may impact previous decisions while altering the basis for future decisions.
- **intercooler**
  a heat exchange device.
- **internal combustion engine**
  a device for converting heat energy into mechanical energy by combusting fuel inside the engine.
- **International System**
  a formal term used for the metric measurement system.
- **inverse-time characteristic**
  the characteristic of an overcurrent protection device whereby it opens the circuit faster as the overload is greater.
- **inyot**
  castings of suitable shape and size intended for subsequent hot working.
- **ion**
  an electrically charged atom or group of atoms.
- **irrigation**
  supply of water to plants for supplementing the quantity of water provided by rain or snow.
- **isometric projection**
  a pictorial drafting technique in which the object being drawn is shown in a three-dimensional perspective.
- **jack**
  1. a device built in the form of a jackstone to retard the velocity, prevent scour, and induce deposition. Wire is frequently strung between the six legs to further reduce velocities and catch floating debris.
  2. a device used for lifting heavy loads.

jack knife bridge
a bridge in which the lifting arms fold on themselves at mid-length when in a raised position.

jack rafter
a short partial rafter which does not reach from the plate to the ridge. It might extend from the plate to a hip rafter or from the valley to the ridge. In special cases, it could be from a valley rafter to a hip rafter.

jacket
an outer casing or cover constructed around a cylinder or pipe, the annular space being filled with a fluid for either cooling, heating, or maintaining the cylinder contents at constant temperature.

jacket water circulation (internal combustion engines)
circulation of water through the jacket between the cylinder lines and barrel.

jacketed pump
a pump equipped with jackets around the cylinders, heads, and stuffing boxes through which steam or other heat may be forced to permit the handling of such materials as pith, resin, and asphalt that are solid when cold but melt upon heating; when pumping materials at high temperatures, cold water may be substituted for steam or heat.

jackhammer or sinker
an air operated percussion-type drill that is small enough for being handled by a man.

jacking
a method of providing an opening for drainage or other purposes underground, by cutting an opening ahead of the pipe and slightly larger than the pipe—at the same time forcing the pipe into the opening by means of horizontal jacks.

Japan wax
a wax like material used in the manufacture of polishes.

Japanning
enamelling applied to a variety of articles, chiefly those made from tin plate. Japan is a varnish made usually with copal and coloured with mineral pigments; the commonest colour, black, is due to asphalt. After varnishing, the articles are stoved, that is, dried in heated chambers; the temperature often attaining 300°F. Several coats are ap-

plied to good work, each being stoved before the next is put on. This results in a durable thick coat.

- **jar**
  in well boring, a device consisting of two long flat links, sliding freely within each other for connecting a percussing drill to the rod or rope which works it so that the drill can be given a sharp blow or jar on the upward stroke to loosen it when jammed in the rock.

- **jar test**
  a simple test to determine the correct chemical dose for a water or sewage, most commonly used for estimating the pH and coagulant requirement for raw water. Various amounts and combinations of chemicals are added to several samples of water of the same volume, placed in jars and stirred slowly.

- **jars**
  in well boring, a connection between the sinker bars, and the poles or cables, made in the form of two links, having a slide on each other of about two feet. The jars permit the tools to fall on the downward stroke, but on the upward jar them, or give them a sharp pull, tending to loosen them from any crevices or cavings that may hold them.

- **jaw**
  holding tool with jaws.

- **jaw chuck**
  a lathe chuck made up of a faceplate provided with adjustable jaws.

- **jaw clutch**
  a positive connection for engaging or disengaging parts similarly to a friction clutch.

- **jaw crusher**
  a primary crusher machine in which motion is imparted to a movable jaw by means of a toggle joint which exerts the maximum force when the jaw is at the extreme of travel on its working stroke.

- **jerk pump**
  a pump which supplies the injectors of diesel engines with the required amount of fuel at a pressure which will open the injector automatically as soon as the fuel in the line has reached the correct pressure level.

- **jet**
  the stream of water issuing at high velocity from an orifice or nozzle or similar opening.

- **jet aeration**
  an activated sludge process using coarse-bubble aeration introduced at about a quarter of the depth of the aeration tank through perforated pipes.

- **jet diffusion basin**
  a stilling basin in which the energy is dissipated by jet diffusion principle to reduce the length of the basin normally required by the hydraulic jump.

- **jet pump**
  a pump where a moving jet of water, steam, or air delivered through restricted throat at high velocity creates a suction and draws in the water to be raised. The velocity energy which is imparted to such water is later converted into pressure energy by an expansion.

- **jetted well**
  a well that is excavated by the use of a high velocity stream of water.

- **jetting**
  a method of well sinking wherein the casing is sunk by driving while the material inside is washed out by a water jet, and carried to the top of the casing.

- **jetting bit**
  a high pressure washing tip for jetting, usually combined with a chisel tip.

- **jetting method of boring**
  a method of boring well in which the material encountered is loosened by a strong current of water directed through a suitable nozzle or bit and is carried to the surface and thus removed by the return stream of water.

- **jetting processes**
  a method of well sinking wherein hollow roads or small pipes are inserted inside the casing. Water is forced out of the lower end of the rods or pipe to wash material through the annular space between the casing and the water pipe to the top of the casing.

- **jetting rig**
  equipment used for drilling holes in the soil by means of a high velocity stream of water.

- **jib barrow**
  a wheel-barrow, a platform without sides.

- **jig drill**
device to hold work while drilling.

- **jig pump**
a portable force pump.

- **jockey arch**
a device used on horse drawn cultivators to hold the two gangs a fixed distance apart. It has the shape of an arch so as to straddle any plants in the row.

- **joint drains**
vertical drains provided in construction joints as a precaution against development of excessive pressure in the joints during the construction period and against seepage of reservoir water through the joints during the operating period.

- **jolt ramming machine**
a molding machine used for all classes of work. The flask rests on a table that is raised by air pressure and then allowed to drop, thus packing the sand.

- **jonval turbine**
an axial-flow turbine.

- **jonval wheel**
a parallel or axial flow water wheel.

- **judicial ditch**
a drainage ditch, whose organisation and operation are functions of a court.

- **jump height curve**
a curve showing relationship between discharge intensity and the subcritical conjugate depth determined for hydraulic jump formation.

- **jump plough**
a plough which is designed to jump over impediments in soil without slowing forward progress.

- **junction box**
a rectangular or round well of permanent material that is installed where two or more tile lines join or where several lines join at different elevations.

- **junction chamber**
a converging section of a conduit used to facilitate the flow from one or more conduits into a main conduit.

- **junction diode**
an electronic device that is made by joining a piece of n-type and a piece of p-type semiconductor material together. In such a device, the current flows only when the p-n junction is forward-biased.

- **junction field-effect transistor (JFET)**
a type of voltage-controlled transistor.

- **junction wells**
wells provided in a pipe under drainage system at changes of grade, or at junction in a main.

- **juvenile water**
water that has never been meteoric water. It comes from deep rock that originally was molten, not sedimentary.

- **k values**
a term applied to denote conductivity of heat of various materials.

- **kalsomine**
a white or coloured wash for the walls of rooms, ceilings, etc.

- **kame**
a geologic term used to designate short ridges of gravel, piled up by a glacial stream flowing from beneath the ice of a glacier, frequently against the end or terminal moraine.

- **kankar**
a nodular variety of limestone which is of spongy nature, containing some quantity of clayey and siliceous matter. It is found either in layers or blocks or in separate nodules and is termed as 'block kankar' or 'nodular kankar' respectively.

- **Kansas dynamometer cart**
a two-wheeled vehicle fitted with hydraulic type measuring devices and recording instruments that is pulled by the drawbar of a tractor and is used to determine implement draft and drawbar pull.

- **Kaplan water turbine**
a water turbine in which the pitch of the blades can be varied in accordance with the load.

- **karman number**
a dimensionless parameter employed in problems in unsteady open channel flow for ensuring that the same type of turbulent flow is obtained in the model as prevails in the prototype. It is equal to the height of equivalent sand roughness times the shear velocity divided by the kinematic viscosity.

- **keeping quality test**
a simple, sample holding test in which a sample of a product is taken to determine the ability to maintain quality in storage.

- **kelly**
  the uppermost joint of drill pipe which permits it to be rotated by the drill and at the same time to move vertically.
- **kellydrive bushing**
  a cylindrical connection or drive with two lugs on opposite sides, between the kelly and the drill.
- **Kessener brush**
  an early method of surface aeration in the activated sludge process, devised by Dr. H. H. Kessener of Holland in 1925, with rapidly rotating stainless steel brushes on a horizontal shaft. The brushes are partly submerged in the mixed liquor, swirled it around and thus aerated. It has been superseded by the TNO rotor.
- **key bolt**
  a plain bolt to keep pieces of machinery together.
- **key drawing**
  a master sheet that contains instruction for the engraver.
- **key wall**
  a cut-off wall extending from the cut-off trench into the bed rock or impervious subsurface stratum.
- **keyhole caliper**
  a caliper having one straight leg and the other curved.
- **kharif crop (India and Pakistan)**
  summer crop or monsoon crop.
- **kicker**
  a vane so placed on an auger shaft as to centrifugally propel materials laterally and outside opening; used to move lint cotton from the chute into the press box.
- **kier liquor**
  the waste liquid from boiling cotton, flax, straw, or other material with a high cellulose content, in a strong alkaline liquid. Kiering removes dirt, grease, colour, etc., thereby allowing the cellulose to be used for papermaking. Cotton yarn also is kiered before dyeing.
- **kilo**
  metric prefix for one thousand (1,000).
- **kilowatt (kW)**
  1,000 watts.
- **kilowatt-hour (kW-hr)**
  commonly used unit of electrical energy use. 1 kilowatt-hour is the equivalent of using 1,000 watts of electrical power for one hour.

- **kilt**
  a local term used in India for a system of working the rope-and-bucket lift.
- **kinetic energy**
  energy of motion.
- **king pin**
  big bolt or pin passing through the centre casting of an engine truck.
- **king post**
  the central upright pieces in a roof truss against which the rafters abut, and that supports the tie beam.
- **king post truss**
  a truss constructed with a king post.
- **king ventilating system**
  a flue ventilating system for barns which introduces fresh air at or near the ceiling and removes foul air from a point near the floor. Usually used where electric power is not available.
- **kit**
  an outfit of tools, accessories etc.
- **knapsack seeder**
  a device for broadcast sowing which consists of a convas sack fastened to a seeding mechanism. A crank, turned by hand, revolves a wheel having radial ribs which throw seeds to the front and sides.
- **knapsack sprayer**
  a device which is carried on the back and shoulders of the operator to which is attached a hand pump to pressurise a liquid for atomisation at the nozzle of the hand gun.
- **kneading machine**
  an appliance to mix, incorporate and knead dough.
- **kneading tools**
  tools for pressing into shape hot gutta percha when covering an insulated joint.
- **knife**
  an instrument made up of a handle attached to a blade with a cutting edge.
- **knife application**
  injection of fertiliser into soil in gaseous form.
- **knife blade fuse**
  a fuse having end connections which resemble the blades of a knife switch and which fit into the contacts of the cut-out in the same way that the blades of a switch fit into the switch contacts.
- **knife clip**
  a steel or malleable fixture used on cutter bars to hold the knife close to the ledger plates.

- **knife coulter**
a rearward curving sharpened blade that is attached to the plough beam so that it cuts the soil just ahead of the point of the plough share.

- **knife coverer**
attached to a planter, which consists of rearward curving blades, which cover seed in the row.

- **knife head**
that end part of the sickle bar of a moving machine to which the pitman is attached.

- **knife section**
one of several truncated, triangle like, sharpened, metal pieces which are attached to the knife bar or the knife of a moving machine.

- **Kubel test**
a test for small amounts of organics in water, which involves boiling the water with potassium permanganate for 10 min. The value is about three times as high as that obtained in 4 ha at 27°C in the permanganate value test.

- **KVA code letter**
code found on an electric motor nameplate. It indicates the approximate kilovolt-amperes (KVA) per horsepower consumed by the motor under locked-rotor condition.

- **labour intensive techniques**
techniques of production which uses relatively more labour than other factors.

- **lacing bars**
bars that connect two leaves of a strut to make them act as one member. The system of bars does not intersect each other between the leaves of the strut.

- **lacustrine peat**
the decomposition of vegetation to form peat, under anaerobic conditions.

- **lacustrine plain**
a plain originally formed as the bed of a lake from which the water has disappeared.

- **lacustrine soil**
soil formed at the bottom of a lake by deposition of mineral particles removed from banks or carried by flowing water; or of debris of organisms.

- **lag (time)**
referring to discharge of water level; the time elapsing between the occurrence of cor-

responding changes in discharge or water level at two points in a river.

- **lagging**
the surface or contact area of a drum or flat pulley especially a detachable surface or one of special composition.

- **lagoon system (manure)**
the manure from some livestock operations caught in trenches and then washed out to a pond called lagoon. Bacteria in the lagoon decompose much of the material and release plant nutrients. Lagoon water makes an effective fertiliser.

- **laid length**
the length measured after pipe is placed in position. The laid length includes such items as gaskets or space between ends of pipe in coupling or the insertion of bell-and spigot joints.

- **lambda**
a unit of volume.

- **lamella separator**
a modern settling device like the inclined-tube settling tank but with parallel, steeply inclined plates, about 50 mm apart, instead of the tubes.

- **laminated**
a soil arranged in very thin layers or plates, horizontal to the soil surface.

- **laminated structure**
an arrangement of the soil mass in very thin plates or layers less than 1 mm. in thickness, lying horizontal or parallel to the soil surface. Usually medium to soft consistency.

- **laminated wood**
an assembled product made up of the layers of wood and adhesive in which the grains or adjacent layers are parallel.

- **land**
a general term to describe the non-water portion of the earth, usually in reference to a large surface area, but it is sometimes used synonymously for soil, e.g., wet lands include many wet soils.

- **land (marginal)**
land from which it is doubtful whether a specified form

of management can produce an income in excess of the costs of such management under given economic conditions, usually with reference to agricultural land.

- **land (sub marginal)**
  1. land which is incapable of sustaining a certain use of ownership status without economic loss.
  2. land which can not be profitably used for agriculture.

- **land appraisal**
  rural land appraisal for determining the value of land as mortgage collateral or for tax assessment. Social land units, mainly farms, are evaluated in terms of potential production within the institutional and legal environment.

- **land arable**
  land which, in its present condition, is physically capable, without further substantial improvement, of producing crops requiring tillage.

- **land capability**
  suitability of land for use, without damage, for crops, grassland, forest or wild life, taking into consideration soil characteristics, physical land features and climate. Land capability involves consideration of (1) risks or hazards of land damage due to erosion or other factors, and (2) difficulties or limitations for different types of land use due to land characteristics and climate. In irrigated areas or areas with adequate well distributed rainfall, climate ceases to be a hazard.

- **land capability unit**
  a group of soils nearly alike in suitability for plant growth and response to soil management. Capability unit provides more detailed information for management of specific fields, than the sub-class.

- **land classification**
  categorisation of land units based on properties of land or its suitability for some specific purposes.

- **land clearing**
  removing of vegetative cover, loose rock tits, and other obstructing materials before land can be prepared for irrigation.

- **land coffer dam**
  an excavation below ground water, often close to valuable existing structures which must be protected and surrounded by limited or very slowly flowing bodies of

water. The surrounding banks are either stable slopes or are maintained by various forms of sheeting and impervious sheeting or may be taken care of by drainage and pumps.

▪ **land crowing**
the process of using earthmoving equipment to elevate the surface of flat land into a series of broad, low ridges known as 'turtle backing' in the sugar cane area. This practice differs from bedding in that the drains each side of the ridge are farther apart, and the elevated area between drains is checked during and after construction to ensure there are no pockets.

▪ **land disturbing activity**
any land change which results in water or wind erosion and movement of sediment.

▪ **land drain, field drain, agricultural drain**
a buried pipe with loose joints, dug into wet farmland to keep it dry enough for cultivation. The drain slopes down either to a ditch or to a bigger drain pipe. It is buried about 1 to 1.5 m deep in peaty soils and about 0.9 m deep in silts.

▪ **land drainage**
the disposal of storm water from rural areas, and the removal of water excess from agricultural lands to prevent or to relieve water logging, the accumulation of harmful amounts of salts and deterioration of soil structure. When specifically referred to agricultural land, it is also known as agricultural drainage.

▪ **land elevation**
the elevation (up-building) of the surface of a flood plain so as to render it less susceptible to flooding.

▪ **land equivalent ratio (LER)**
a criteria used to evaluate the efficiency of an intercrop system which takes into account the amount of land occupied by a component crop under mixed cropping. LER of a component is defined as the ratio of yield per unit area of a given species in an intercrop system to its yield per unit area as a sole crop.

▪ **land evaporation**
soil evaporation plus interception.

▪ **land forms**
physical features of the land surface developed by the pro-

cesses of erosion and sedimentation, such as gullies and larger valleys, residual upland divides, glacial troughs, wave-cut cliffs and wind blowouts, etc.

- **land grading**
  reshaping of the field surface to a desired elevation and slope. It is necessary in making a suitable field surface to control the flow of water, to check soil erosion and to provide surface drainage.

- **land holding (economic)**
  a land holding which allows a man a chance of producing sufficient to support himself and his family in reasonable comfort after paying his necessary expenses.

- **land holding (operational)**
  that area of land which is actually cultivated by the farmer and his family. It includes land kept fallow.

- **land leveller**
  a piece of equipment that removes the high areas and fills the low spots of a field so as to produce a surface with a continuous smooth slope.

- **land levelling**
  a process of shaping the land surface for better movement of water and machinery over the land.

- **land management**
  all activities of man that affect the land; as a flood control technique it includes conservation practices on agricultural land, grazing, regulation, forest utilisation on a sustained yield basis.

- **land phase of flood**
  the initial phase in flood generation when water running over the surface of the ground collects in tiny rills to evacuate eventually into the stream channels.

- **land plane**
  a large, wheeled machine used to level the land.

- **land planning**
  tillage operation that cuts and moves small layers of soil to provide a smooth and refined surface.

- **land reclamation**
  making land capable of more intensive use by changing its character or environment or both through operations requiring collective effort. Does not include the clearing of stumps, brush, and stones from land, or simple techniques of erosion control that can be effected by the individual.

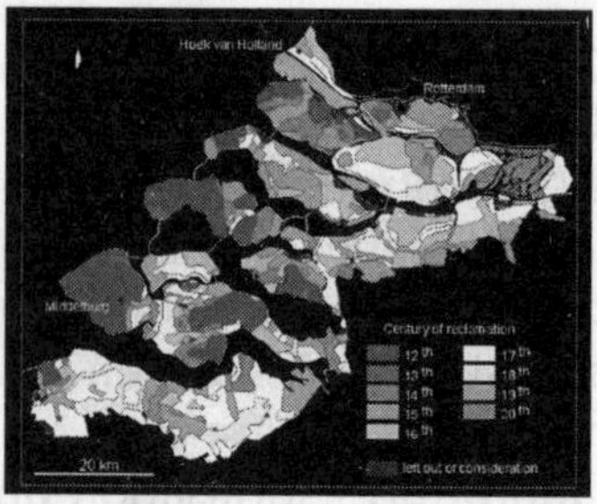

- **land resource area**
  an area of land reasonably alike in relation to agriculture with emphasis on problems in soil and water conservation.
- **land resource region**
  a generalised grouping of land resource areas having regional relationships to agriculture with emphasis on problems in soil and water conservation.
- **land resource unit**
  a sub-division of land resources area with emphasis on specialised agriculture and soil and water conservation problems.
- **land resting**
  temporary stoppage of cultivation on a piece of land.
- **land roller**
  an implement, of a number of different kinds which is used primarily for crushing colds and compacting or firming the soil in the preparation of seed beds.

- **land sat**
  satellites that take photographs of the earth and plot the earth's resources.
- **land slide**
  1. rapid movement downslope of a mass of soil, rock, and debris.
  2. mass of material that has slipped downhill under the influence of gravity, many times assisted by water (when the material is saturated). Rapid movement of a soil mass, rock or debris down a slope also constitutes landslide.
- **land smoothing**
  the operation of producing a plane land surface with a continuous slope.

- **land subsidence**
the general lowering in elevation of a considerable area of land surface, due to the removal of the underlying supporting material either by artificial causes, such as underground mining operations, or removal of soluble material by means of water.

- **land topography**
the surface configuration of the land.

- **land treatment**
any method of treating sewage or sewage effluent by irrigation.

- **land type**
land uniformly possessed of particular distinguishing characteristics. A natural land type is land having a particular set of defined natural characteristics, principally of soil, climate, relief, stoniness, and native vegetation.

- **land use**
the culture of the land surface which has a determining effect on the broad social and economic conditions of a region and which determines the amount and character of the runoff and erosion. Three general classes are recognised: Crop, pasture or range, and forest.

- **land use capability**
an interpretative grouping of soils based on the inherent soil characteristics, external land features and environmental factor which limit the use of land.

- **land use capability classes**
division of agricultural land into classes of similar production potential with a view to making the best use of each piece of land without causing excessive erosion, or loss of productivity.

- **land use classification**
the classification of land according to the use to which it is put.

- **land use map**
a land use map tends to evaluate the lands actually used on the one hand and also the lands available for further use on the other. The map should be helpful to predict the use and misuse of land to recommend its suitability for better use like farming, forestry, pasture growing, settlement etc. under different systems of management. Hence, the ultimate purpose of land utilisation

survey and map is to arrive at systems of land use and management suitable the kinds of resources and capabilities of the land composing it.

- **land use pattern**
  the area design or arrangement of land uses, major and minor, and of operation units convenient for cultivation.
- **land use plan**
  plan showing recommended location and intensity of development of a piece of land for habitation, recreation, agriculture, plantation etc.
- **land use planning**
  a development of plans for the uses of land that will, over a long period, best serve the general welfare, together with the formulation of ways and means of achieving such uses.
- **land utilisation index**
  has been defined as the number of days during which the crops occupy the land during a year divided by 365 and can be expressed as a fraction or as a percentage.
- **land wheel**
  the wheel that travels on the unploughed soil on landside of any type of plough.
- **land, super marginal**
  land that shows profit after meeting all expenses on its use.
- **land, wild**
  uncultivated land which may or may not be maintained by owner for productive vegetative growth.
- **landforming**
  tillage operation which move soil to create desired soil configurations. Forming may be done on a large scale such as contouring or terracing or on a small scale, such as ridging or pitting.
- **landing**
  1. a horizontal platform between flights or stairs or at the top or bottom of a flight of stairs.
  2. a place where logs are assembled for transporlation in loads or rafts.
- **landing of the beam**
  the angle that the plough beam makes with the line of the landside. It varies with the size of the plough. Wedges are forced between the beam and the frog, or landside to obtain the desired landing.
- **landing place**
  a horizontal platform near the water surface level where the

cattle or traffic make a landing when unloaded out of the canal.

- **landing the point of the share**
the projection of the point of the share toward the unploughed land by an amount of 1/8 to 1/4 inch.

- **landlocked**
enclosed by land, or nearly enclosed, as a harbour (thus, protected from the sea).

- **landmark**
a conspicuous object, natural or man-made, located near or on land, which aids in fixing the position of an observer. Such an object in the sea is known as a seamark.

- **landmark fishway**
a type of pool and jet fishway formed by making the baffles perpendicular to the bottom, but oblique to the axis, and by placing a simple jet deflector close to the opening left by the baffles. Named after its designer Landmark (Norway).

- **lands**
part of the piston that provides support to the piston rings.

- **landscape**
natural land features such as hills, forests, fields, water etc. which distinguish one part of the earth's surface from another. The landscape is usually confined to an area, which can be comprehended by the eye in a single view.

- **landscape design**
the profession concerned with the planning and planting of outdoor space to secure the most desirable relationship between land forms, architecture, and plants to best meet human needs for function and beauty.

- **landscape evaluation**
an attempt to judge the quality of the landscape.

- **landside**
a sidepiece opposite the plough moldboard, forming a V with the share edge, as in bar share, and which guides the plough and receives the side pressure when the furrow is turned. Not all ploughs have it.

- **landside clearance**
the adjusted space of 1/2 to 3/4 inch-between the heel of the landside on the rear bottom of

a trailing plough, and the plane of the furrow sole. The rear wheel J is adjusted to provide this clearance. On all moldboard plough mounted on wheels it will be noticed that the heel of the landside does not touch the floor when properly set. The vertical suction in this case will be the amount the heel of the landside is elevated above the floor.

- **landslide spring**
  a spring occurring at the base of a landslide.

- **landweeds**
  weeds growing on lands, i.e., on cropped lands and on banks and berms of irrigation and drainage channels.

- **lane's weighted creep theory**
  the premise of this theory is that the horizontal contacts with the foundation and slopes flatter than 45°, being less liable to have intimate contact, are given only one-third the weightage of steeper (45° and greater than 45°) and vertical contacts while calculating the creep length. The creep length obtained by the above procedure is called lane's weighted creep length. The hydraulic gradient is assumed constant along the Lane's weighted creep length.

- **langelier saturation index (symbol i)**
  an indication of the non-corrosiveness of a water, measured by the over saturation or under saturation of calcium carbonate in it. It is defined as the difference between the pH of the water under consideration and its pH when in equilibrium with calcium carbonate ($CaCO_3$) (the saturated pH or pHs). A negative value of I implies under saturated water that will dissolve $CaCO_3$ and will not deposit a protective film of $CaCO_3$ on the pipe of stop corrosion. A positive value of I implies that $CaCO_3$ will be deposited on the pipe. The pHs value can be calculated by measuring the alkalinity, calcium ion concentration, total dissolved solids and temperature of the water. The chalk test is not essential. The calculations are described in Standard Methods for the Examination of Water and Waste Water.

- **lantern**
  in a centrifugal pump, a hollow casing on the engine side of the pump body.

- **lap joint**
  a joint made by one piece lapping over another and connected by a fastener, welding, or glue.

- **lapse**
  in meteorology the change of temperature, usually cooling, of air as its height increases.

- **lapse rate**
  the rate at which air temperature changes with increasing height.

- **lard press**
  a mechanical device consisting of a hopper box and a screw press which is used for pressing warm lard from the cooked fat of pork.

- **large plots**
  plots from 1/200 to l/100th hectare in size for the comparative testing of varieties with the use of farmers seeding and harvesting methods under ordinary farm conditions.

- **larros**
  hard subsoil of ferruginous appearance.

- **laser**
  a device that produces coherent light.

- **latasol (laterite) soils**
  it has been recommended that the use of the term 'laterite' be discontinued as defined in 'Soils and Men'. The term 'latasol' has been suggested as a replacement to refer to the more friable and iron-rich soils of the hot, moist, or wet-dry tropics referred to as 'red-loams'.

- **latasolization**
  the process by which latasols are produced; that is, soils having the following characteristics: (1) a low silica-sesquioxide ratio in the clay fraction, (2) low activity of clay with medium to low cation-exchange capacity of the mineral fraction, (3) low content of all but most resistant primary minerals, (4) low content of soluble material, (5) high degree of aggregate stability, and (6) red colour of reddish shades of other colours. Horizons of concentration of iron and aluminium oxides may be present as a result of removal of silica.

■ **latent heat**
the heat released or absorbed, per unit mass, by a system during a change of phase. In meteorology, at O°C the latent heat of vaporisation, fusion and sublimation of water are about 630,80 and 680 cal/g. respectively.

■ **latent heat flux density (e)**
energy used for evaporation; a mechanism of dissipating the radiation absorbed by the surface.

■ **latent heat of fusion**
the quantity of heat required, to change a body from the solid to the liquid state.

■ **latent heat of vaporisation**
the quantity of heat required to change a liquid to a vapour.

■ **latent heat of water**
denned as the quantity of heat required to convert one gram of ice at its freezing point into water at the same temperature. The value of latent heat of water is found to be 80 calories. This is also referred to as latent Heat of Fusion of Ice.

■ **lateral**
1. secondary or side channel.
2. a branch sewer buried in the street, collecting sewage from the house drains, and delivering into an intercepting sewer.

■ **lateral bund**
in soil conservation, an embankment constructed at intervals along the contour bund, and constructed along the slope of the land and at right lines to the contour bund, to prevent flow towards and accumulation of water, in a depression especially when there is deviation from contour.

■ **lateral canal**
1. a separate navigational canal constructed parallel to the general course of a river which does not lend itself to canalisation.
2.a channel leaving or entering the main channel from the side; a branch channel.

■ **Law of Charles and Gay-Lussac**
states that the volume of a gas varies directly with its absolute temperature, provided the pressure remains the same.

■ **Law of Conservation of Energy**
states that energy can neither be created nor de-

stroyed, although it can be transformed from one kind to another.

- **L-beam**
  a steel beam with a cross section in the shape of a letter I.

- **light-emitting diode (LED)**
  a special type of diode that emits light when forward-biased.

- **line graph**
  a type of graph showing the relationship between two continuous variables.

- **linear actuator**
  a hydraulic device that produces straight-line motion, such as a hydraulic cylinder.

- **linear foot**
  a length of 12 inches, without regard to thickness or width.

- **live load**
  a load on a structure that changes under normal building use.

- **load**
  any electrical device that consumes power, such as a lamp or motor.

- **lobe profile**
  refers to the shape of the camshaft lobe.

- **lower plumb cut**
  the cut made on the lower end of a rafter.

- **lubrication system**
  engine auxiliary system that keeps engine parts lubricated.

- **lysimeter**
  an enclosed volume of soil and plants arranged in a tank so that they can be weighed periodically or continuously. The lysimeter is in the open so that the rainfall, runoff and groundwater flow can be measured and the evapotranspiration determined. The soil and crop conditions in the lysimeters should be close to the natural conditions.

- **m**
  a symbol indicating strong cementation. It is applied as a suffix to horizon designations to indicate irreversible cementations.

- **maccrator**
  a screenings disintegrator. A garbage grinder.

- **machete**
  a very large, heavy knife which is used for cutting sugarcane and brush.

- **machine banded pipe**
  a pipe made of wooden staves, held together in a machine and

tightly wrapped with wire. It is made in definite lengths in a factory and joined in the field by couplings.

- **machine grading**
movement of earth by bulldozers or other equipment to establish a predetermined grade.

- **machining**
a cold metal working process for shaping metal.

- **macrobiota**
the larger organisms in soil, including plant roots, larger insects and earthworms.

- **macropores**
soil pores, which are big enough to facilitate rapid downward movement of water by force of gravity; where soil is sandy or well granulated.

- **macrostructure**
soil structure presenting the aggregates of the mineralogical complexes into larger units, as the result of cementation.

- **made land**
land filled with earth or mixed earth and trash under the control of man. A miscellaneous land type.

- **mafic**
a general term used to describe ferromagnesian minerals.

- **magnetic induction**
the process of producing current flow in a conductor by the relative motion of the conductor and a magnetic field.

- **magneto-type ignition system**
a type of engine ignition system that produces electricity for the ignition spark by magnetic induction.

- **main**
an important pipe, often one carrying water. If buried, it should have at least 0.9 m of earth cover as frost protection.

- **main bearings**
the supports for the principal shaft that supplies or transfers power in a machine are called main bearings.

- **main body**
the structural part of a dam above foundations.

- **main canal**
the main channel or conduit taking off from the source of supply, from which branches or main laterals, and distributaries receive their supplies. Normally, no direct irrigation is permitted from the main canal.

▪ **main channel**
the middle, deepest, or best navigable channel. This term is applied to water boundaries in sounds, bays, straits, gulfs, estuaries, and other arms of the sea—also to lakes and landlocked seas in which there are deep-water sailing channels.

▪ **main crop**
the principal or most important crop grown in a farm or area.

▪ **main disconnect**
the over-current protection tool/switch that limits amperage and allows all electrical service provided through a main or building service entrance panel to be disconnected.

▪ **main drain**
a principal drain which serves as an outlet to all subsidiary drains which discharge into it.

▪ **main duct**
a tunnel which distributes or collects air to or from either two or more smaller (lateral) ducts or to a perforated or slatted floor; used in connection with conditioning of harvested produce.

▪ **main journal**
crankshaft bearing surfaces that hold the crankshaft in proper alignment within the engine block.

▪ **main line**
in a sprinkler irrigation system, the conduit conveying water under pressure from the source of supply to feed submains or sprinkler laterals. The supply line, feeding the giant sprinkler system, is often referred to as a lateral.

▪ **main line meter**
a water meter installed on a large main of the distribution system.

▪ **main spillway**
the spillway which is designed to pass the spillway design flood and which usually comes into operation. The main spillway may be assisted by emergency spillway in passing the design flood.

▪ **main water table**
the upper limit of the zone of saturation above the first impermeable or almost impermeable formation.

▪ **main wheel**
a large wheel equipped with lugs on a binder or harvester which bears a high percentage of the weight of the machine and which furnishes the power to operate the machine.

maintenance
the operations performed in preserving irrigation or drainage canals, hydraulic structures, service roads, and works in good or near original condition without increasing its/their capital costs. Repairs are part of maintenance.

maintenance cost
the cost of maintaining a structure, flood control works, a canal system, drainage system, network of service roads, including expenses of staff and other overhead expenses. The maintenance cost for comparison, is expressed in case of a reservoir per unit of storage capacity; in case of a canal system per unit of discharge at head, or peg unit of area irrigated, or per unit length of the canal; in case of flood control works per unit of area protected; and in case of drainage channels per unit of area protected or per unit length of drainage channel.

maintenance interval
manufacturer's recommendation for periodic machine service. It is given as hours of use and/or length of time.

maintenance period
the period after the completion of the construction of work during which the contractor is responsible for making good the repairs or damage that may occur to the work during its operation. The contractor is entitled to final payment and/or refund of security only after the expiry of this period.

maize shelter
an equipment to remove maize grain from cobs.

major estimate
a term applied to the estimate for a work, when sanctioned amount of the works expenditure exceeds a specified amount. This term is also applied, for the sake of convenience, to the work itself.

major project
a project which can irrigate more than 10,000 ha.

make up water
the proportion of the feed water that is needed to make up the losses of water in a boiler system.

male
that one of a pair of corresponding parts which fits into the other; as, a shaft in a bearing or a bolt in a nut, the other piece being termed the female element; external threads are known as male threads.

man land ratio
the ratio or relationship between numbers of human being and land which may exist at any given time for a given geographical area.

management, watershed
the administration and regulation of the aggregate resources of a drainage basin for the conservation of water and the control of erosion, stream-flow and floods.

manifold
a pipe fitting with numerous branches used to convey fluids between a large pipe and several smaller pipes.

manifold chamber
a plenum chamber serving more than one duct; used in connection with conditioning of harvested produce.

manifold vacuum
the vacuum in the intake manifold that develops as a result of the vacuum in the cylinders on their intake strokes.

manometer
an instrument for measuring the pressure of fluids; a pressure gauge or vacuum gauge. The pressure of the fluid is balanced against a column of mercury, or other standard fluid is balanced against fluids in a U-tube or against the elastic force of a spring, an elastic diaphragm, or the like.

mantle
the unconsolidated organic and inorganic materials together with more highly disintegrated bedrock, found on land surfaces as the result of geological weathering.

manure application
process of spreading manure on the soil and in some cases mixing it with soil.

manure bucket
a scoop attachment for a tractor which is used to lift and load manure.

manure fork
1. a sturdy fork which is used for handling manure.
2. the attachment for a loader used to lift and load manure.

manure spreader
a wagon-type implement which is used for carrying barnyard manure to the field, shredding it and spreading it uniformly on the land. The power for the spreading mechanism and conveyor is supplied from the rear wheels

or from a tractor power take-off.

- **manuring**
mixing of bulky manures like farmyard manure and compost with the soil before sowing or planting. Using manure spreader, country plough, shovel tooth cultivator, a blade harrow or a disc harrow does this.

- **marble**
a general term for any limestone that will take a good polish. Strictly, a metamorphosed and recrystallised limestone. Named according to the place of quarrying (Makarana marble, Halian marble), chemical composition (magnesium marble, calc-silicate marble), colour (green marble, grey marble), appearance (mottled marble, veined marble). Marble test.

- **marbled**
the presence of two or more distinct colours in approximately equal amounts not blended but more or less mixed in occurrence in the soil mass. In a marbled soil there is no general or predominant colour, as in the case of a mottled soil.

- **marginal bund**
a backwater bund; a term used in India and Europe for an embankment built along a river upstream of a weir at a short distance from the margin.

- **marginal farmer**
a farmer with a holding of size not more than one hectare of irrigated land. In case of other areas this limit is raised taking into account the availability of irrigation facilities, the soil type, the cropping pattern and other relevant factors.

- **marginal land**
land that is sufficiently fertile to yield a return that just covers the costs of production. Marginal land will be abandoned if the price obtained for its products falls or the cost of production increases.

- **marine deposit**
a deposit laid down in ocean water.

- **marine material**
material deposited in the waters of oceans and seas and exposed by elevation of the land or the lowering of the water level.

- **marine pollution**
the sea is polluted in three main ways: (1) By raw or imper-

fectly treated sewage both from the land and from ships and by rubbish from ships. Ships are now able to have activated sludge plants for which the activated sludge may be supplied dry. (2) By trade effluents, often discharged untreated. (3) By oil or oil-derived materials such as tar from petroleum refineries and similar works on the coast, as well as from ships, since 60 per cent of the world's vast oil output is carried over water. Nutrients from pollution of types (1) and (2) can cause coastal areas to become eutrophic, especially if there is little flushing action by coastal currents. The chief immediate victims of oil pollution are diving sea birds. Their plumage becomes coated with oil and spoilt. They cannot avoid swallowing oil when they try to preen themselves clean and it may poison them.

- **marine soil**
soil formed from materials deposited by the waters of oceans or seas, exposed by elevation of the land or desiccation of the sea.

- **marker system**
a system with signal lights set on the rim of the dial scales of the batchers to mark the desired weight of each material for each class of concrete.

- **market gardening**
the growing of an assortment of vegetables for local or roadside market.

- **marks feed**
on a fertiliser drill, an inverted cone with a stationary projecting lip beneath which a revolving plate brings the fertiliser to the lip which deflects it to the centre of the cone and drops it through the spout to the soil.

- **marl**
an indefinite term applied to any earthy, crumbling deposit containing varying quantities of calcium carbonate, clay soil, sand, and carbonaceous material.

- **marly**
limnic material that consist of marl.

- **marsh**
periodically wet or continually flooded areas with the surface not deeply submerged. Dampness is due to impermeable nature of the soil and poor drainage. Vegetation consists of hydrophytes like sedges (Cirrus spp.) cattails (Typha spp.) and rushes etc. Sub-classes are fresh water

marsh and marine marsh. The term is often used synonymously with swamp.

- **marsh hay**
a dry vegetative material harvested in low areas and sometimes used to cover drain tile joints in organic soils.

- **marsh, tidal**
a low, flat area having interlacing channels and tidal sloughs, being periodically inundated by high tides. Vegetation is characterised by salt tolerant plants.

- **mashie spur**
a streamline spur of L-shape made of bushing, with a shank at right angle to the bank and arm turned downstream with generous curvature, and the arm flared away from the axis to the channel with 1 in 10 slope.

- **masonry check**
ditch barriers built of rock and mortar with the surface flush, and conforming to the cross-section of the ditch.

- **masonry cum earth dam**
a dam comprising portion of earth and masonry along its length (length of spillway portion being excluded).

- **masonry dam**
a dam constructed of stone set in mortar, or of concrete blocks. The adjective masonry may be qualified as: Rubble masonry, cyclopean masonry, and concrete masonry.

- **masonry lining**
a type of canal lining involving the use of bricks, tiles, concrete blocks and stones. According to the construction used, they are separately called brick lining,Tile Lining, Block Lining and Stone Lining.

- **masonry paint**
a paint applied to masonry units to prevent water seepage through the units and also used for decoration.

- **masonry reservoir**
a reservoir with masonry walls instead of earth embankments, or with masonry lining.

- **masonry syphon**
a syphon, barrel/barrels of which is/are built of masonry (bricks or plain concrete).

- **mass**
quantity of matter in an object.

- **mass balance**
matter is neither created nor destroyed; therefore in any treatment process: physical, chemical or biological, whatever goes in must come out.

- **mass centre**
the centre of gravity of the mass curve.

- **mass concrete**
concrete produced with relatively small amount of mortar to lower down the heat generated and the rise in temperature and to reduce the shrinkage, etc., and to improve durability. It has little or no reinforcement against shrinkage and temperature cracks. Used for massive construction.

- **mass curve**
a curve in which values of cumulative rainfall or runoff are plotted against time.

- **mass diagram**
a diagram, curve, or graph, plotted with rectangular co-ordinates, representing a summation (integration) of all preceding quantities up to a point, each ordinate being equal to the sum of preceding terms in the series, with the corresponding abscissa representing elapsed time, etc. The diagram is used extensively in storage and regulation studies pertaining to stream flow.

- **mass discharge curve**
a curve with accumulated values of recorded discharges plotted against time intervals. The average slope of the curve represents average discharge.

- **mass flow**
a liquid exposed to a greater pressure in one region moves from the region of higher to that of lower pressure and this phenomenon is called mass flow or bulk flow. It differs from diffusion flow is being proportional to the pressure gradient instead of concentration gradient.

- **mass movement**
unit movement of a portion of the land surface, as in creep, landslide, earth flow or subsidence.

- **mass point**
a mechanical mode of a vibrating system consisting of vibrator and a certain part of the soil upon which the vibrator rests. Hence, the mass point rests on an elastic support. A mass point is imagined as containing the properties of a body such as weight; e.g., the centre of gravity of a body is a mass point taken as an idealised equivalent of that body.

- **mass rainfall curve**
a plot of accumulated precipitation against time.

- **mass transport**
in regard to surface gravity waves (progressive) the nonperiodic movement of water in the direction of wave travel.

- **mass water content**
the water content expressed as the weight of water in a soil divided by the oven-dry weight of soil. When expressed as a percentage, it is the mass water percentage or simply water percentage.

- **massive**
a large mass of soil manifesting irregular shape and cleavage.

- **massive soil structure**
large, uniform masses of cohesive soil which sometimes have irregular cleavage, as the C horizons of many class soils.

- **mast frame**
a frame in which a tower or vertical beam carries one or more load lines at its top.

- **master clutch**
a clutch which transmits all power from the engine and controls both travel and the power take-off.

- **matc**
maximum acceptable toxicant concentration.

- **match marking**
a system of marking the parts or members of a structure so that they always may be connected in exactly the same order and manner.

- **material records**
records relating to all materials used on the site, date of arrival, the period of store, the date of permanent location and the dates when requisi-

tions were made for their supply.

- **materials handling**
any operation which changes the location of a material without changing its form or make up except as an incidental effect.

- **mathematics**
the branch of science that deals with quantities, magnitudes and their relationships through the use of numbers and symbols.

- **matric potential**
the amount of work an infinitesimal quantity of water in the soil can do as it moves from the soil to a pool of free water of the same composition and at the same location. This work is less than zero or negative work, thus reported in negative values. Matric potential nearly equals water potential in non-salty soils.

- **matrix**
natural material having larger particles embedded in it.

- **mattress**
a blanket of burhs or poles interwoven or otherwise lashed together and placed to cover an area subject to scour, weighted with rock, concrete blocks, layers of stones as well as blanket of bush or poles, etc., or otherwise held in place. Also the section of a bank revetment below normal low water.

- **maturation pond**
an aerobic waste stabilisation pond, usually following a facultative pond, typically 1.5 to 3 m deep. Two maturation ponds in series, each with a detention period of 7 days, are needed after facultative ponds to reduce the final; effluent BOD_5 to 25 mg/litre. Algae are fewer in maturation ponds than in facultative ponds and even fish may be present in the last maturation pond.

- **mature river**
a stream whose slope is so reduced that the water velocities are just sufficient to carry the detritus delivered by the tributaries, which have steeper slopes than the main stream. When all channels, including wet weather rills are graded, the stream system is said to be old. Old streams have a

wide flood plain, a broad meander belt, and gentle slopes.

- **mature shore line**
  a shore line which in its horizontal form, or in plan, has become relatively stable as far as wave erosion is concerned.

- **mature soil**
  a steady stage of its parent material adjusted to the environment. The time factor has no significance after the soil reaches maturity.

- **mature stream**
  a graded stream: One which is in balance;. A stream whose energy is just sufficient to transport its load. It neither silts nor erodes.

- **mature valley**
  a stream valley which has reached a stage of development where cutting of the bottom has practically ceased. Such valleys have a flat bottom, with numerous tributaries of considerable length, and a medium flat gradient.

- **matured charges**
  the portion of a repayment contract represented by the instalments due, whether paid or not.

- **maturing**
  1. ripening
  2. of a trickling filter, the process of acquiring a balanced population of bacteria, etc., to purify the sewage passing through it. A new trickling filter may take 6 months to mature in a cold winter but only a few weeks in warm weather. Early summer is therefore the best time to start a trickling filter. The term is also used of an activated sludge process that is starting, but mature sludge can be introduced from another aeration tank very easily.

- **maximum acceptable toxicant concentration (matc)**
  the highest observed concentration in air or water of a pollutant which has no effect on reproduction, growth, behaviour,, etc., of a plant or animal. For air it has been superseded by threshold limit values so far as humans are concerned.

- **maximum and minimum thermometer**
  a thermometer designed to record the maximum and minimum temperatures during a period of time. It usually consists of a graduated capillary tube at the base of which is a bulb containing

ethanol. Within the capillary is a thin thread of mercury with a steel index at each end. When the temperature rises, the alcohol expands, thus forcing the mercury up the tube, carrying the upper steel index. Therefore the position of the upper index marks the maximum temperature reached, and similarly the position of the lower index, the minimum temperature attained.

- **McCormick, Cyrus**
patented and manufactured the horse-drawn reaper.

- **mechanical drawing**
a drawing made with the aid of mechanical drafting tools, like T-squares, triangles, etc.

- **mechanical efficiency**
the ratio of the power delivered by a machine to the power produced by the machine.

- **mechanical property**
feature of a metal including hardness, ductility, toughness, tensile strength, and compression strength.

- **melter**
a machine used for melting the caps removed from honeycomb by the uncapping knife or plane. The melting process makes it possible to separate the wax from the honey in the caps.

- **melting point**
the temperature at which a matter changes from a solid to a liquid.

- **messenger**
the neutral wire conductor in a triplex cable.

- **metal**
any of a class of chemical elements that has ductility, malleability, lustre, and can conduct heat and electricity.

- **metal inert gas welding (MIG)**
a kind of gas metal arc welding using a continuous wire as the electrode and an inert gas to shield the molten puddle.

- **metallic conduit**
metal tubing used to support and protect a number of individual electrical wires which are run through it.

- **metallurgy**
the study of the physical and chemical properties of metals.

- **metric system**
the international system of measurement based on the gram, metre, and litre as the basic units of weight, length, and capacity, respectively.

▪ **milk replacer**
dried vegetable or dairy products that are mixed in warm water and fed to calves in place of milk.

▪ **milli**
metric prefix meaning one one-thousandth (001).

▪ **milli-ampere (mA)**
quantity of current flow equal to one one-thousandth (001) of an ampere.

▪ **model number**
manufacturer's identification number for a specific product, such as electric motors or engines.

▪ **motor**
a device for converting an energy input into a mechanical energy output.

▪ **motor controller**
a switch or other device that is normally used to start and stop a motor by making and breaking the electrical current to the motor.

▪ **motor duty cycle**
an electric motor rating indicating whether the motor is designed for continuous or intermittent operation.

▪ **motor rotation**
the direction of rotation of the motor shaft.

▪ **multi-meter**
a type of electrical meter that can be used to measure two or more electrical circuit characteristics.

▪ **multiplication of force**
the transformation of a small input force into a larger output force, accompanied by a proportional decrease in output distance or speed.

▪ **munsell code**
a soil colour code published in Baltimore Maryland. A soil colour is expressed by a symbol formed of letters and figures.

▪ **mutant**
in biology, an individual with different characteristic from the rest of its species, which may be transmitted to its descendants. Pesticides kill all the pests except those mutants that can resist them, and often the pests develop a strain that is immune to the pesticide.

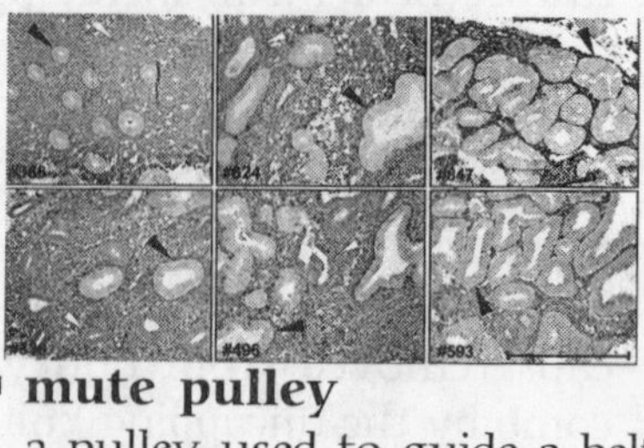

▪ **mute pulley**
a pulley used to guide a belt and which is adjustable to

various positions upon its stand.

- **mutual interference of piles**
when two cut-offs or pile-lines are placed closely in the permeable soil below a hydraulic structure, a change is caused in the ultimate value of uplift pressures due to overlapping effect of stream-lines, which is known as mutual interference of piles. It takes note of the depth of either cut-off, the distance between them and the floor length.

- **nahrhumus**
the material in humus, consisting of carbohydrate, organic acids, protein and soluble or hydrolysable substances, which is readily decomposed by micro-organisms.

- **nailing groove**
steel framing which has a groove through its midsection to permit nailing of wood rafters on such.

- **nameplate**
tag located on the frame of an electric motor that provides specialised information about the motor.

- **nappe**
the sheet or curtain of water overflowing a weir or dam. When freely overflowing any given structure it has 4 well-defined upper and lower surface.

- **narrow base terrace**
a terrace similar to a broad-based terrace in all respects excepting the width of ridge and channel; the base width of a narrow terrace is usually from 4ft to 8ft.

- **narrow based ridge terrace**
type of contour terrace for soil conservation. Commonly called bund, and consists of a comparatively narrow-based embankment constructed at intervals across slope of land either on a level (along the contour) or slightly graded longitudinally, if conditions require removal of excess water.

- **narrow crested inclined weir**
a weir formed by projecting an inclined wall (upstream or downstream) from the bed of the channel. Briefly, these are called inclined weirs.

- **narrow track seeder**
a seed drill which has the wheel closely spaced leaving the ends

of the seed box projecting beyond them on either side.

- **National Electrical Code (NEC)**
comprehensive set of electrical safety standards published by the National Fire Protection Association and updated after every three years.

- **National Fire Protection Association (NFPA)**
a non-profit organisation devoted for promoting and improving the science and methods of fire protection. It publishes the *National Electrical Code.*

- **native**
occurring in a region at the time of its discovery: ıdigenous.

- **native nutrients**
nutrients present in the soil itself.

- **natric horizon**
illuvial horizon of silicate clay accumulation, over 15 per cent exchangeable sodium and columnar or prismatic structure.

- **natural**
pertaining to nature. Characteristics of objects that have not been modified or affected by the application of human intelligence.

- **natural bed (stone)**
the surface of a stone on which it was originally deposited, but not necessarily horizontal when resting in the quarry, the strata being often inclined and even upright, due to the beds having become folded by volcanic and other agencies.

- **natural blanket**
a cover of naturally occurring soil acting as a blanket. The natural blanket is generally less permeable.

- **natural bridge**
a bridge or arch of natural rock, which is sometimes formed when the tough, un-jointed rocks of a valley spur are cut through by a stream. The phenomenon is more common in limestone country.

- **natural combustion**
combustion of organic substances on the surface of the soil or in the soil, with the intervention of microorganisms. It is favoured by light and heat and slowed down by drought or an excess of water.

- **natural draft ventilation**
a ventilating system using natural forces (wind, natural convection) to force air

through a building or product.

- **natural drainage**
as applied to a given area, refers to the natural condition of that area for disposing of surface water from it.

- **natural draught**
the height of a chimney and the warmth of the air in it cause the air to be sucked up from the fire below. The higher the chimney and the warmer the air, the greater is the draught. An electrically driven fan, however, is much cheaper than a high chimney. Forced-draught fans and induced draught fans are now used.

- **natural dry density (ndd)**
dry density of soil in its undisturbed state.

- **natural ecosystem**
an ecosystem that ultimately produces the climax community, which develops in the absence of any major disturbance and consists of a complex, self regulating food web of several tropic levels.

- **natural erosion**
wearing away of earth's surface by water, wind, ice, gravity etc. in natural environment of climate and vegetation without the interference of man.

- **natural flow**
the flow of a stream as it occurs under natural as opposed to regulated conditions.

- **natural frequency (self frequency)**
frequency for free vibration of a system.

- **natural levee**
a low, alluvial ridge adjoining the channel of a stream, composed of sediment deposited by flood water which has overflowed the banks of the channel. It is common to streams that flow through alluvial valley land.

- **natural load**
the amount of sediment carried by a stable stream.

- **natural runoff**
it is the runoff from an area which will discharge without artificial works intended to collect the waters which fall upon it.

- **natural soils**
the thin top layer of the earth's crust supporting vegetation, formed largely as a result of natural (geological, climatic,

chemical and biological) processes.

- **natural strainer well**
a well which taps a water bearing formation of imperfectly assorted and at least partly incoherent material, and whose specific capacity has been increased by the withdrawal of much of the fine grained part of such material in the immediate vicinity of the screened portion of the well.

- **natural subirrigation**
the delivery of water of plant roots by the capillary fringe.

- **natural subsurface irrigation**
application of irrigation water below surface of soil. Possible where an impervious layer exists below the root zone, where water is led to a series of ditches dug up to the impervious layer, which then moves laterally and wets the root zone, idea being to raise the table so that plant roots can get water by capillary movement.

- **natural surface line**
line joining points of the natural ground surface parallel to or along the centre line of the bed of a canal.

- **natural system**
a system of drains, similar to the trunk system, but located along the swales, depressions, or low spots to conform to topography.

- **natural water**
water as it occurs in its natural state usually containing other solid, liquid, or gaseous materials in solution or suspension.

- **natural water table**
a water table in its natural condition and position, not disturbed by artificial additions or extractions of water.

- **natural watercourse**
a surface or underground water course created by natural agencies and conditions.

- **natural well**
an abrupt depression in the land surface, not made by human agency, which extends into the zone of saturation but from which water does not flow to the surface except by artificial process. It is distinguished from such features as ponds, swamps, lakes and other bodies of impounded surface water which also extend into the zone of saturation in general by having a smaller water surface, being

deeper in proportion to its water surface area, and having steeper sides. While it is not a well in a strict sense, the name has been firmly established as applying to such a feature.

- **naturally screened well**
  a well which taps an aquifer or imperfectly assorted and at least partly incoherent material, and whose specific capacity has been increased by the removal of the fine material by use of proper developing equipment through a screen.

- **navigation canals**
  artificial canals primarily meant for transport by water.

- **nazzaz**
  a compact, impermeable pan, concretionary in character, occurring at a slight depth below the surface of red sandy soils in the Levant.

- **neanthes**
  a genus of polychaetes, some species of which may be used in bioassays of tidal mud and estuaries.

- **neck**
  a narrow stretch of land connecting two larger areas or a narrow body or channel of water between two larger bodies of water.

- **necton, nekton**
  tiny organisms that swim strongly, unlike plankton and neuston.

- **needle**
  1. a piece of timber or steel forming an abutment for the end of raking shores; or a timber or steel beam laid horizontally at right angle to the wall and supported on both sides by dead shores.
  2. a timber set on stream bed or weir crest or regulating gate to close in opening for the control of water; it may be either vertical or inclined; a form of stop plank.
  3. a closing member as in needle valves.

- **needle bearing**
  a form of roller bearing employing a large number of small diameter rollers without a cage or spacers; well suited to applications involving oscillating motion.

- **needle dam**
  a movable dam composed of upright pieces or needles, the lower ends of which bear against a sill on the river bot-

tom, the upper ends of which bear against some form of bridge.

■ **needle pitman**

a connecting rod on the binding attachment of a grain harvesting machine or a baler.

■ **needle points**

1. needle-like pins used to faten fine cabinet work together; as dowels.
2. needles used as points for compasses, dividers, and other mathematical instruments used for engineering drawings. The needles are usually locked into the instrument with a nut and bolt so as to be easily renewable.
3. a short length of wood or steel placed in the face of a wall to take the abutment of shores.
4. sharp projections on a beam used to hold dead shores in position for the support of the upper part of a wall during underpinning.

■ **needle valve**

the term 'needle valve' is usually applied to a valve having a circular orifice which is closed by a conical plunger closing in the direction of flow. It is used for regulating purposes, necessitates operation of the valve for long periods of time at any opening from fully closed to fully open.

■ **needle weir**

a movable timber weir in which the wooden barrier consists of upright square timbers placed side by side against the structural-steel frames.

■ **negative area**

an area where deposition is taking place.

■ **negative artesian head**

an artesian head which is at a lower elevation than that of the surrounding water table.

■ **negative confining bed**

a confining bed that prevents or retards downward movement of ground water where the overlying water has sufficient head to produce a resultant downward pressure.

■ **negative feedback**

tendency of a system of counteract externally imposed change and return to a stable state.

■ **negative lead**

in valve gears, the amount by which the steam port is closed to admission when the piston is at the bottom of the cylinder.

- **negative pressure**
  a pressure less than the local atmospheric pressure at a given point.

- **neon timing light**
  a light that is used for the accurate adjustment of spark timing in magneto or battery ignition systems. It is connected in series with No. 1 spark plug, and the flashing light makes visible a timing mark on the engine flywheel.

- **neosol**
  a soil of recent formation.

- **nephelometer**
  a type of turbidimeter that measures turbidity through the amount of scattering of the light that falls on turbid water.

- **nephelometry**
  the measurement of light scattered by particulate matter in suspension. It is commonly used for the detection and quantification of unwanted particles or haze in liquids and is a valuable technique in certain types of quantitative chemical analysis, e.g., determination of sulphate. It is more sensitive than tubidimetry. One very useful additional advantage is that traces of haze may be accurately measured in highly coloured fluids.

- **neptunian land**
  sedimentary land supposed to have been deposited formerly on the bottom of a sea.

- **neritic zone**
  the sea bed between low water and the edge of the continental shelf.

- **nernst and zeta potential**
  the adsorption of ions on the surface of the colloidal particle and a diffused zone of oppositely charged ions in the dispersion medium leads to a potential difference. But when electrophoresis takes place a portion of these oppositely charged ions adhere to the colloidal particle and there is therefore a movement not between the actual surface of the colloidal particle and the main body of the liquid but between the layer adhering to the colloid particle and the main liquid. The total drop in potential between the actual surface of the colloidal particle and the liquid medium is spoken as the nernst potential while in fall of potential between the liquid adhering to the particle and the liquid medium is known as electro

kinetic potential or zeta potential.

- **Nernst lamp**
an incandescent lamp having for its light giving element a pencil composed of the refractory oxides of rare earth, termed the 'glower' which becomes incandescent upon the passage of an electric current. The glower is an insulator when cold, but becomes a conductor when heated. The necessary heat is conveyed to the glower by means of 'heaters' consisting of thin porcelain tubes wound with find platinum wire and coated with a refractory paste. A cut out device opens the circuit through the heater when the glower arrives at the proper temperature, and a steadying resistance called the 'ballast' prevents the glower's burning out.

- **nesslerisation**
a measurement of the ammonia content of water by the formation of a yellow colour when it is added to Nessler's reagent—a solution of potassium mercuric iodide with potassium hydroxide.

- **net above ground productivity (nap)**
accumulation of biomass in above ground parts of plants (trunks, branches, leaves, flowers, and fruits) over a specified period; usually expressed on an annual basis.

- **net appreciation on livestock**
the amount by which the value of the livestock on hand at the end of the year plus the amount received for livestock sold exceeded the livestock on hand at the beginning of the year plus the amount paid for the livestock purchased.

- **net area sown**
the actual magnitude of land which was cultivated during the year, the area sown more than once being counted only once. Thus, the net area sown may be equal to or less than the actual geographical area of the plot. However, the area sown more than once is added to the net area sown, to arrive at the total area sown during the year.

- **net assets**
the working capital of a business or company, i.e., the difference between current assets and current liabilities.

- **net available head**
the difference in elevation between the last free water surface of the water in a power conduit above the water wheel and the first free water surface in the conduit below the water wheel.

- **neutral bar**
connection point in the service entrance panel that is intentionally connected to earth.

- **neutron**
the part of the atomic nucleus having no electrical charge.

- **Newbold, Charles**
he obtained the first US patent for the plough.

- **nominal voltage**
standard voltage classification which may differ slightly from the actual or measured voltage. The most common are 120V or 240V.

- **non-differential cylinder**
a double-acting, linear hydraulic actuator having both sides of the piston fitted with a cylinder rod.

- **non-ferrous**
a metal that does not contain iron (Fe).

- **non-metallic conduit**
non-metal tubing used to support and protect a number of individual electrical wires which are run through it.

- **non-positive displacement**
a type of pump where the fluid volume delivered per cycle decreases with increases in outlet pressure.

- **normalising**
a treatment process of metal heat where the metal is heated to just above the transformation temperature, maintained for a time, and then allowed to cool in open air.

- **nursery**
a place where young trees, shrubs, and other plants are grown.

- **nyirok soils**
red soils of Moravia and Hungary, with affinities to a terrarosa, but may have resulted from earth tertiary weathering.

- **oat kicker**
a machine which is used to remove weed seeds from wheat.

- **ob**
a prefix meaning reversed, turned out.

- **obelisk**
an upright four sided pillar, gradually tapering as it rises and terminating in a pyramid.

- **obligate aerobe**
a microbe that can live only in an environment containing oxygen, unlike a facultative anaerobe.

- **obligate anaerobe**
a microbe that can live only in the absence of free oxygen, unlike a facultative anaerobe.

- **oblique projection**
a mechanical drawing showing three sides of an object, with the near side drawn at a 90-degree angle.

- **oblique weir**
a weir the crest of which is inclined to the flow in plan.

- **obsequent**
flowing in a direction opposite to that of the dip of the strata or the tilt of the surface; said of some streams and contrasted with consequent.

- **obsequent stream**
a stream that flows in a direction opposite to the dip of the formations and opposite to the direction of flow of the original consequent stream.

- **observable variable or observations**
a variable, the values of which can be directly observed. The basic facts of any statistical investigation are called OBSERVATIONS.

- **observation error**
the difference between an observed value of a quantity and a value adapted as representing the ideal or true value of that quantity.

- **observation pipes**
pipes installed at any desired location and to any desired depth in a basin, etc., for the measurement of ground water level.

- **observation points**
the points at segmented intervals along a discharge section line at which the velocities and depths are measured.

- **observation well**
hole made in the soil upto a certain depth for observing water table or piezometric level.

- **observed value**
a value of a quantity that is obtained by instrumental measurement of the quantity.

▪ **occupation crossing**
crossing (bridge or culvert) which gives access between two portions of a farmer's land.

▪ **Occupational Safety and Health Administration (OSHA)**
agency of the US Government responsible for formulating and enforcing workplace safety standards.

▪ **occupier's water rate**
water rate charged on the area actually irrigated and which usually varies with the nature of crop or crops, the quantity of water normally required by the crop, the scarcity or abundance of water at the time of supply, and the value of the crop.

▪ **ochric epipedon**
a light coloured surface horizon generally low in organic matter which includes the eluvial layers near the surface.

▪ **octane number**
standardised rating of the ignition quality of gasoline.

▪ **odd area**
a small area like fence corner or an irregularly shaped area etc. which may serve as wild life habitat.

▪ **odometer**
an instrument attached to the hub of a wheeled vehicle to measure distance travelled.

▪ **off season**
the season of the year when a particular crop is not normally grown or cannot be grown.

▪ **off season tillage**
any tillage that is carried out between two crop periods.

▪ **offset**
in plumbing: (1) a combination of elbows or bends which brings one section of a line of pipe out of line with, but into a line parallel with, another section, (2) and a pipe fitting in the approximate form of a reverse curve to accomplish the same purpose.

▪ **offset disk harrow**
a harrow adapted for use in orchards and vineyards which inherently runs to one side of the tractor centreline and thus can be used, to cultivate under branches too low for the tractor to pass under. It consists of one right hand gang

and a left hand gang operating in tandem.

- **offshore terrace**
a deposit of sand which is built out into deep water by the combined action of waves and currents. It trends parallel to the shore. The materials are dumped over its outer edge. Its upper surface is a continuation of the shore bench or wave cut bench with which it is associated. Shore terraces are not exposed except by sufficient subsidence of the water level.

- **offstream storage pond**
reservoir adjacent to a continuously or intermittent flowing stream, with an intake, either a pipe or open channel, that diverts water from the stream into the pond.

- **offtake canal**
a canal for taking away water from another canal.

- **ohm**
the basic unit of electrical resistance.

- **ohmmeter**
a device used to measure electrical resistance.

- **Ohm's Law**
basic electrical law stating the relationship between voltage, amperage, and resistance in a circuit.

- **ohvine**
orthosilicate of iron and magnesium, crystallising in the orthorhombic system, and occurring widely in the darker basic and ultra basic igneous rocks.

- **oil coalescer**
a device for separating tiny oil particles from water.

- **oil control rings**
the lower ring or rings on a piston designed to prevent excessive amounts of oil from working up into the combustion chamber.

- **oil filter**
that part of the lubricating system that removes dirt and dust from the oil circulated through it.

- **oil level indicator**
the indicator, usually a stick, that can be removed to determine the level of oil in the crankcase.

- **oil pressure indicator**
oil gauge that reports to the driver the oil pressure in the engine lubricating system.

- **oil ring**
bottom piston ring(s) that are intended to control the amount of oil on the cylinder wall.

- **oil stone**
a block of abrasive material bonded together, used for removing metal.

- **old fallows**
cultivable lands not cultivated for a number of years.

- **old stream**
a river which is graded along with all its tributaries.

- **oligotrophic**
description of a water with little life because of the low content of nutrients.

- **on costs**
costs made up by listing the actual cost to the contractor of all the preliminaries and ancillary works, such as the transport, erection and dismantling of all site buildings and of all plants and materials.

- **open circuit**
a circuit condition where a complete circuit from the source to the load and back to the source does not exist.

- **operating system**
engine system required in order for the primary engine system to operate.

- **orthographic projection**
a type of mechanical drawing that shows two or three separate views of the object drawn.

- **osasis effect**
effect of dry fallow surrounds on the microclimate of a relatively small acreage of land, where an air mass moving into an irrigated area will give up sensible heat. For small field, this may result in a higher ET crop as compared to predicted ET crop using a climatic data collected inside the irrigated area; conversely ET crop predictions based on weather data collected outside the irrigated fields may overpredict actual evapotranspiration losses.

- **osmosis**
movement of materials through a semi permeable membrane.

- **output device**
computer hardware used for transferring computerised information onto paper or other media. For example, printer, monitor, etc.

- **over-current**
level of current flow in a circuit above the maximum safe level.

- **overhead valve cylinder head**
type of cylinder head housing the intake and exhaust valves.

- **p**
a sub-horizon symbol for ploughing or other disturbance. The symbol p is used as a suffix with A to indicate disturbance by cultivation or pasturing. Even though a soil has been truncated and the plough in or is clearly in what was once B horizon, the designation A_p is used. When an A_p is subdivided, the Arabic number suffixes follow, as A_{pl} and A_{p2}, for the A_p is considered comparable to A^ A2 or B_2.

- **pace on pace factor**
the length of an average step, which is approximately 3 ft for most persons. A stride is two steps.

- **pack aquifer ratio**
the ratio of gravel to sand in a gravel envelope used with well screens.

- **package job contract**
a contract in which the government or organisation undertaking a construction project makes only one party responsible for all services connected with the work, including planning, design, drawing of specifications and construction. This type of contract may be made either on a firm price or on cost plus basis. It is based on a functional type of specification, describing the general features and requirements of the finished plant. The contractor draws up all the detailed engineering designs and specifications, and performs the construction and testing. The work is considered complete when he turns over the plant in satisfactory operation to the owner.

- **packaging**
the process of filling, weighing and sewing bags.

- **packed tower**
an early type of wet scrubber, consisting of a hollow tower packed with media down which the reagent (scrubbing or wash water) trickles. Packed towers can have upward (counter current) gas flow, downward (concurrent) flow or cross flow.

- **packer**
in well drilling, a device lowered in the lining tubes, which swells automatically, or can be used to expand by manipulation from the surface at the correct time, to produce a watertight joint against the sides of the bore hole or the casing, thus entirely exclud-

ing water from higher horizons.

- **packer arms**
arms on grain or corn binding equipment which pack the material against the trip arm before the tying mechanism. They are curved arms which oscillate and pick up the material and force it against the trip arm which is spring loaded.

- **packer grouting**
the use of expandable implements to isolate sections of a grout hole and thereby enabling the application of grout to a portion of the hole only.

- **packer shaft**
the power shaft which runs the packer arms on a binder.

- **packing coefficient**
a parameter in sediment transportation theory denned as d^2 times the number of grains per unit of area, when 'd' is the diameter of the grain.

- **packing hook**
a tool used to remove packing from stuffing boxes.

- **packing house**
1. a separate building or shed adjacent to a range of greenhouses where flowers and greenhouse plants may be packed for market,
2. a building which is equipped and arranged for grading and packing fresh fruit or vegetables.

- **packing shed**
a roofed but otherwise open structure in which fresh fruits, vegetables, etc., are sorted, packaged and graded for the market.

- **pad filter**
a form of filter press which is used in the wine and brewing industries for the final or polishing filtration of wine and beer.

- **paddle**
one of the broad boards, or flats, at the circumference of a water wheel.

- **paddle wheel**
a simple device operated by persons treading on foot treadles, with arms and upper part of the body supported on a bamboo frame for the watering and dewatering of paddy and other fields.

- **paddle wheel aeration**
the mechanical agitation of sewage in the aeration tanks of the activated sludge process by means of paddle

wheels. Paddock. Small fenced field used for grazing purposes.

- **paddling**
  the process of using a stick to firmly tap clay in order to shape it, also used to modify the surface texture clay.

- **paddy shelter or dehusker**
  an equipment to remove husk from paddy grain.

- **paddy weeder**
  an equipment for interculture used in paddy cultivation. It is used for uprooting weeds and burying them in puddled soil between rows of standing crop. It improves the aeration of the soil.

- **painter's putty**
  a mixture to whiting and linseed oil, sometimes including whitelead, forming a plastic substance.

- **paira crop (uthera)**
  a crop sown broadcast in the standing crop of low land rice before its harvest where the residual moisture is used for the establishment of uthera crop, e.g., lathy-rus, gram, lentil etc. in standing crop of rice.

- **paired comparison**
  the comparison of two samples of equal size where the members of one can be paired off against the members of the other.

- **paired obstacle fishway**
  a fishway which consists of a sloping rectangular channel with baffles or obstacles in opposed pairs, spaced along the channel at intervals approximately equal to the channel width, and with straight free passage between them.

- **pairie type combine**
  a type of combine or harvester thresher for small grain, either tractor drawn or self-propelled, which is primarily adapted for level land as opposed to the hillside-type combine. Prairie-type combines are usually characterised by a smaller straw rack

area per ft. of cutter bar width than newer combines used in humid regions.

- **paiudization**
process regarded by some workers as geogenic rather than pedogenic, including the accumulation of deep (>30 cm) deposits of organic matter as in mucks and peats.

- **pakihi soil**
a soil developed under podsolic conditions and impeded drainage. Found in parts of New Zealand.

- **palaeolimnology**
the study of past limnology from the records provided by sediment cores extracted from a lake bed.

- **paleosol**
soil which formed formerly, under conditions generally different from those of the present day.

- **paling fence**
a light, modern fence which is made of narrow slats or pales, sharpened and driven into the ground a few inches apart to be supported at top and bottom by horizontal strips nailed to each pale. It is used for gardens, backyards, and small enclosures.

- **pallet**
a platform which has an upper and usually a lower deck spaced and held together by wooden stringers. It is lifted and moved by the forks of the lift equipment.

- **pallets**
wood frame structures which are used to support a stack of bags, to prevent them from touching the floor and absorbing moisture.

- **pan**
pan is a layer formed by accumulation of materials such as salts, clay, etc., and which impedes free drainage.

- **pan formation**
a relatively thin and hard layer or stratum of more or less impervious material occurring in the subsoil, and frequently extending over a considerable area, usually occasioned by the accumulation of salts of iron, calcium, magnesium and other bases where there is no free drainage through the soil.

- **pan method**
a method of grading strawberries, which consists of emptying each box into a grading pan and removing the defec-

tive berries as the fruit is poured back into a box.

- **pan, pressure or induced**
  a subsurface horizon or soil layer having a high bulk density and a lower total porosity than the soil directly above or below it as a result of pressure by normal tillage operations or by other artificial means. Frequently referred to as plough pan, plough-sole, tillage pan, or traffic pan.

- **panel**
  a compartment, a distinct portion, or a section of shuttering; usually required to be reused.

- **panel door**
  a door made up of panels with alternate rails and stiles between.

- **pannier**
  a conical basket, with a pole passing through its axis, used as an anchor in bridge or pontoon laying.

- **pans**
  horizons or layers in soils that are strongly compacted, indurated or have high clay content.

- **pans in soils**
  horizons or layers in soils that are strongly compacted, indurated, or very high in clay content are often called pans.

- **pantograph**
  an instrument for enlarging or reducing a map.

- **paper sleeves**
  tubes of paper used for covering the projecting end of reinforcement bars at the joints to prevent the surrounding concrete from bonding to the bars.

- **parabolic weir**
  a measuring weir whose notch is bounded on the sides by parabolas.

- **paradox control**
  an automatic mechanical follow-up valve, which controls water pressure to ensure positive positioning of a needle (closing member).

- **paradox gate**
  a high pressure service gate used as a companion to the ring follower gate for opening or closing off flow under heads up to and in excess of 600 feet (182.9 metres).

- **parallel canal**
  a canal built by the side of, and parallel to, the parent canal (main canal, branch canal or lateral) to irrigate areas not permitted to be irrigated from the parent canal.

parallel circuit
electrical circuit having alternate paths for current flow.

parallel cropping
a type of multiple cropping in which two crops of dissimilar growth habits are grown simultaneously in such a way that they do not affect the performance of each other adversely.

parallel drainage pattern
drainage pattern in which streams or their tributaries are parallel or nearly parallel to each other.

parallel field ditch system
a system similar to bedding except that the drains are spaced farther apart and have a greater capacity than the dead furrows.

parallel flow
the flow of 2 fluids in the same direction.

parallel lateral open ditch system
a system similar to field ditch system, except that the ditches are deeper and cannot be crossed with most machinery.

parallel link hitch
a type of hitch used extensively on mounted row crop cultivators.

parallel shift
a method of control designed to prevent cultivators from ploughing up plants in which the gangs are rigidly connected together at the front by bars.

parallel system
a system of drains with long parallel laterals emptying into a single main drain.

parallel terrace system
a terrace system where adjacent terraces are parallel or nearly parallel and the point row area is reduced or eliminated.

parallel wire stretcher
a form of lineman's wire clamp for gripping a wire and bringing it to the proper tension.

parapet
a low barrier (wall, vertical posts and horizontal rails, etc.) at the two edges of the bridge roadway and also on the top of abutment wings to protect the traffic and the pedestrians.

parent canal
with reference to any canal, the canal which feeds it with water supply.

▪ **parent material (soils)**
the unconsolidated, chemically weathered mineral or organic matter from which the A and B horizons (solum) of soils may have developed by pedogenic processes. The C horizon may or may not consist of materials similar to those from which the A and B horizons developed.

▪ **parquet flooring**
thin pieces of hardwood laid in some kind of a pattern on a wood subfloor, usually laid in glue and secured with panel pins. Parquet flooring is thinner, cheaper and more decorative than wood blocks but is less durable.

▪ **parshal flume**
a calibrated device for measuring the flow of water in open conduits. It consists of a contracting length, a throat and an expanding length.

▪ **partial analysis**
an ingenious device invented for the purpose of treating economic phenomena in relatively small sectors of the economy. It takes into account only a limited number of variables for intensive study and investigations.

▪ **partial association**
the measure of association in sub-populations of qualitative characteristics analogous to the partial correlation between quantitative variates.

▪ **partial correlation**
a mean of determining the relationship between two variables when the effect of other variables is eliminated.

▪ **partial cutoff**
a cutoff extending into the underlying stratum, but not reaching an impervious subsurface stratum.

▪ **partial duration series**
a list of all events, such as floods, occurring above a selected base, without regard to the number within a given period. In the case of floods, the selected base is usually equal to the lowest annual flood, in order to include at least one flood each year.

▪ **partial performance**
this term conveys the idea of performance which, while sufficient to confer appreciable benefit on the recipient, falls short of the concept of substantial, or virtually complete, performance.

partial regression
in regression analysis, the coefficient of an independent variable in the complete regression equation, that is to say, the regression equation involving all the variables under consideration.

partial vacuum
the description of a space condition in which the pressure is less than atmospheric, but in which the vacuum is not complete.

partially developed underground water basin
an underground water basin wherefrom the use of underground water does not exceed the naturally occurring recharge but a considerable development, either agricultural or urban, dependent on the use of underground water has taken place on the land above.

partially drowned weir
a weir at which the level of water upstream is being affected to some extent by the level of water downstream.

partially penetrating artesian well
an artesian well which does not reach the bottom of the confined aquifer.

partially penetrating well
a well which is less than fully penetrating.

partially separate system
a sewerage system in which each street has two sets of sewers, one foul sewer and one for rainwater. Unlike the rigidly separate system, the rainwater from the backyard and house back roof drains into the foul sewer.

partially submerged jet
a jet whose central line is below the tail water level, but not top far below to cause the jet to be drowned out by the tail water during operation.

partially submerged orifice
an orifice with a partially free jet.

partially suppressed contraction
1. a reduction in the contraction resulting from the approach walls being so close to the sides of the opening that the full magnitude of the contraction is not obtained.
2. partial or complete suppression of the contraction on one or more sides, the remain-

ing sides exhibiting some degree of contraction.

- **particle board**
manufactured wood product made by bonding wood chips with synthetic resins.

- **particle density (real density)**
real density or the particle density is the density of oven-dry soil particles when air is excluded.

- **particle size**
the effective diameter of a particle measured by sedimentation, sieving, or micrometric methods.

- **particle size analysis**
the determination of the various amounts of the different separates in a soil sample, usually by sedimentation, sieving, micrometry or a combination of these methods.

- **particle size distribution**
the proportions of grains of different sizes that make up a particular sand, gravel, refuse, etc., usually expressed in percentage by weight.

- **particular equilibrium**
an equilibrium, when it pertains to a single variable.

- **Pascal's Law**
basic law of hydraulics, stating that the pressure applied to an enclosed fluid is transmitted equally and undiminished in all directions, acts with equal force on equal areas, and acts at right angles to the walls of the container.

- **pass**
1. a navigable channel, especially at a river's mouth.
2. a narrow connecting channel between two bodies of water; an inlet.
3. an opening through a barrier reef or sand bar.
4. a defile through rough terrain.

- **passive absorption**
absorption by freely transpiring plants occurring through anesthetized or dead roots, or even in the absence of roots.

- **passive absorption of water**
the absorption of water by the roots as a result of forces originating in the leaves such

as transpiration pull and does not require any expenditure of metabolic energy.

- **passive earth pressure**
the maximum value of earth pressure. This condition exists when a soil mass is compressed sufficiently to cause its internal shearing resistance along a potential failure surface to be completely mobilised.
- **passive equilibrium**
plastic equilibrium obtained by a compression of a mass.
- **passive state of plastic equilibrium**
plastic equilibrium obtained by compression of a mass.
- **passive transport**
the movement of substances across a plasma membrane which does not involve the expenditure of energy by the cell.
- **paste wood filler**
material composed of ground silica and quick-drying varnish, thinned to brushing consistency with mineral spirits or turpentine prior to application, worked into the pores of the wood to obtain a smooth finish.
- **pasteurisation**
a process of holding a food product at a certain temperature for a given time in order to kill all pathogenic bacteria and reduce the total microbiological population.
- **pasture**
an area meant for the production of native or introduced forage, which is grazed.
- **pasture cutter**
a stalk cutter with a larger wheel and more knives than a regular tractor stalk cutter for cutting weeds and sprouts.
- **pasture improvement**
any practice meant for improving vegetation for grazing, such as contour furrowing, liming fertiliser application, scattering of droppings, seeding, manuring or other method of management.
- **pasture land**
an area of land covered with grass or other herbaceous forage plants used for grazing animals.

- **pasture management**
application of practices meant to encourage active growth of

pasture plants for providing palatable feed of high nutritive value, while crowding out weeds, brush or inferior grass spp.

- **pasture renovation**
  improvement of a pasture by tillage, seeding, fertilization and some times by liming also.

- **pasturing**
  regular and controlled grazing by cattle of vegetation growing on banks and berms of irrigation and drainage channels. It is a method of weed control.

- **patching**
  repairing or restoring small isolated surface area.

- **paternoster pump**
  a chain pump, named from a fancied resemblance of the discs and the endless chain to a rosary.

- **path lines**
  the traces made by a single particle over a period of time.

- **paved invert pipe**
  generally used in connection with a relatively smooth bituminous material which completely fills the corrugations, intended to give resistance to scour and corrosion.

- **pavement bricks**
  bricks made of useful for paving purposes.

- **pavement pumping**
  ejection of soil and water mixtures from joints, cracks and edges of rigid pavements, under the action of traffic.

- **pavers**
  men who lay or set paving stones or paving bricks; also concrete mixers.

- **paving**
  a cover of stone or other material used to protect an earth dam face, a dike, or shore from the erosive action of waves and currents.

- **payloader**
  basically a tractor that has a hydraulic front end lift, may be mounted on wheels or on crawlers.

- **pea harvester**
  a mechanical device consisting of a winged drum revolving rapidly over a stationary or sharp edged plate which knocks the pods back into a platform from which they are elevated into a box or bag. Green peas are also harvested with a regular mowing machine and the cut vines are taken to a viner for hulling.

- **pea lifters**
lifters which consist of guards attached over the regular guards, which project to the front quite a distance so that pea vines can be lifted, allowing the cutter bar to slide underneath and cut off the stems below the heads.

- **peak**
a maximum quantity or height which occurs over a relatively short period of time.

- **peak demand**
the maximum monetary load placed (1) upon an electric generating plant or system or (2) upon a water plant or pumping station.

- **peak discharge**
the highest stage or discharge attained in a flood.

- **peak discharge from a watershed (QP)**
the highest discharge recorded or reported during the interval of a defined event, expressed in cubic feet per second. It is intended to express the relative magnitude of a flood as to the effects it can produce, that is impart damages or sediment production.

- **peak flood**
maximum rate of flow that occurred during a flood event.

- **peak flow**
the maximum rate of flow which is recorded at a gauging station.

- **peak inflow**
the rate of inflow during floods when the reservoir gains in storage at the fastest rate.

- **peak load**
the maximum load consumed or produced by a unit or group of units in a stated period of time.

- **peak load plant**
an electric power generating plant used primarily to generate power to supply peaks in the load which occur on the system.

- **peak load station**
a generating station which is operated to provide peaking capacity.

- **peak or maximum supply (v max)**
average daily supply requirement during the peak water use period for given crop or cropping pattern and climate; m^3/day.

- **peak period consumptive use**
  the average daily evapotranspiration rate of a crop at its maximum during the period between normal irrigations.
- **peak stage (reservoir)**
  stage reached during floods in a reservoir when the flood inflow has passed its peak and its rate has declined to a figure equal to the rate of outflow.
- **peak supply period**
  water use period for a given crop or cropping pattern during the month or period thereof of highest water requirements; mm/day.
- **peak unit flow**
  peak of a given unit hydrograph.
- **peanut digger**
  any of various plough like devices for removing the peanut and the vine from the soil. This may vary from a one-horse turning plough, with moldboard removed, to a machine potato digger, etc.
- **peat**
  partially decomposed plant material that accumulates in waterlogged anaerobic conditions in temperate humid climates, often forming a layer several metres deep. Peat varies from a light spongy material to a dense brown humidified material in the lower layers. It is used as a fuel and is the first step in coal formation.
- **peat bank**
  a place from which peat is cut.

- **peat clay**
  the lowest layer of a moor in which mineral and organic constituents are mixed.
- **peat hag**
  ground from which peat has been dug or cut.
- **peat humus**
  an organic parent material originating from hypnum moss or reed-sedge peat in a state of advanced decomposition.
- **peat lump**
  a weak layer of original peat found in reclaimed pools. When dry, it is stone hard and compact.

- **peat soil**
composed predominantly of organic material highly fibrous, with easily recognised plant remains.

- **peat winning**
cutting or digging of peat from the soil for fuel.

- **peatery**
a peat bog or bank.

- **peatmoss**
an excellent mulch. Widely used to help 'lighten' a heavy soil. Care must be taken to prevent drying out entirely for then the mulch absorbs water away from plant roots.

- **peatship**
the status of peat.

- **pebble**
a small wind or water worn stone which is larger than gravel but smaller than cobble. Pebbles have a diameter of 10-50 mm. Large number of pebbles that accumulate in one place form a pebble bed.

- **pebble bed**
see pebble.

- **pebble bed clarifier, banks filter**
a rectangular, upward flow, secondary sedimentation tank with a layer of pea gravel hung near the outlet. The effluent is forced to flow up through the gravel before it passes over the outlet weir. The 150 mm deep gravel bed acts as a coarse filter after the humus sludge has settled. The top of the gravel is 150 mm below the top water level. Periodically the water level in the tank is lowered so that the gravel layer can be hosed down and cleaned. It is most suitable for polishing the effluent from a small sewage works.

- **peck**
a measure of volume equal to one-fourth of a bushel or eight quarts dry measure. A pocket of decayed wood in the living tree usually associated with cypress and incense cedar.

- **peck guard**
any kind of material installed on wall board inside a poultry house to keep the birds from pecking at the material and pulling it apart.

- **pectography**
  the microscopic study of the appearance of deposits left on a partition after evaporation of a colloidal solution.

- **pedal**
  a lever or machine part actuated by the foot, as the pedals of a bicycle by which it is driven; the pedals of a tractor or motor car which actuate various controls, etc.

- **pedalfer**
  a soil in which there is a zone of aluminium and iron oxide accumulation in the profile but with no horizon of carbonate accumulation.

- **pedestal**
  a structural member located at predetermined position connecting the sill beam to the capping slab for uniform distribution of load in the case of isolated capping slab.

- **pediment**
  a uniformly sloped (graded) rock surface cut by erosion, such as a mountain slope, and with only a shallow cover of alluvium.

- **pedocal**
  a soil with a horizon of accumulated carbonates in the soil profile.

- **pedochemical weathering**
  the disintegration and chemical modification of minerals taking place within the soil A- and B-horizons, with all the associated biological and other soil forming processes.

- **pedode**
  the time necessary and sufficient for the establishment of the dynamics of a soil.

- **pedogenesis**
  caused by the natural processes of soil development. Synonyms; soil genesis, soil development and soil formation.

- **pedological features**
  are units distinguishable from the enclosing or enclosed materials for any region such as origin, difference in concentration or arrangement.

- **pedology**
  a branch of soil science which deals with the elucidation of the natural laws governing the origin, formation and distribution of soils. Pedology should not be confused with soil science, which is a broader subject, embracing soil fertility, soil technology, soil reclamation, soil physics, chemistry and soil biology or

rather the physics, chemistry and biology of soils.

- **pedon**
the smallest volume that can be called a soil. It has three dimensions. It extends downward to the depth of plant roots or to the lower limit of the genetic soil horizons. Its lateral cross-section is roughly hexagonal and ranges from 1 to 10 sq m in size, depending on the variability in the horizons.

- **pedosphere**
the part of the earth in which soil forming processes occur.

- **pedotubule**
a micropedological features consisting of soil material (skeleton grains or skeleton grain plus plasma, as distinct from concentrations of fractions of the plasma) and having a tubular external form, either single tubes or branching systems of tubes; its external boundaries are relatively sharp. Tubular form, in this context, means that the feature as a unit, or its impression in the enclosing soil material, has a relatively uniform cross-sectional size and shape, most commonly circular or elliptical; that is, the impression of the pedotubule conforms to the definition of channels.

- **pedoturbation**
biologic, physical churning and cycling of soil materials, thereby homogenising the solum in varying degrees.

- **peeler**
1. a rotary veneer machine that cuts logs into thin sheets.
2. a device which removes peels from fruit or vegetables.

- **peg**
a piece to be driven into the ground to mark a boundary, or to hold a rope.

- **pegging**
the jamming of a particle in a screen opening (like a peg in a hole). It can be reduced by using wedge wire screens.

- **pegging out**
marking out the area to indicate the desired position for individual plants or trees.

- **pelagic**
concerned with the open sea, not shallow water.

- **pellet application**
a method of application of manures and fertilisers, by making them into pellets with clay or cow dung and placing them near the root zone.

- **pelleted seed**
  layer of particulate matter around the seed. The pelleting is usually made from finally pulverised limestone or mineral phosphate. Use of synthetic or arabic gum helps in holding the layer of these materials on the surface of the seed. Seed pelleting is some times practised following inoculation of legume seeds with appropriate culture, to protect the rhizobia on seed from the toxic effects due to hazards of pesticides and fertiliser application or due to soil alkali or acidity. Pelleting involves addition of gum arabic or carboxy methyl cellulose to the inoculant slurry before application to the seed. Powdered $CaCO_3$ or rock phosphate is then added to freshly inoculated wet seeds in a suitable container and mixed rapidly for 2 to 3 minutes until the seeds are evenly coated. These seeds are allowed to harden by spreading on clean surface. Pelleted seeds can be stored for 2–3 weeks at about 4°C. Powdered gypsum, dolomite, superphosphate, basic slag and bentonite have also been used as pelleting materials.

- **pelletiser, densifier**
  a machine that is fed with pulverised waste derived fuel, which it extrudes as a solid cylinder of 10 mm or so diameter, broken off in convenient lengths or pellets.

- **pelletized products**
  loose material in a solid state which has been formed into small masses of similar size and shape, usually cylindrical or oval in shape.

- **pellicular front**
  the even front, developed only in pervious granular material, on which pellicular water depleted by evaporation, transpiration, or chemical action is regenerated by influent seepage.

- **pellicular water**
  water held against the force of gravity in the zone of aeration between the soil water zone and the capillary fringe.

- **pellicular zone**
  the maximum depth from the natural surface up to which the evaporation can have its effects.

- **pelton turbine**
  an impulse turbine (usually mounted on a horizontal shaft and with curved vanes or

buckets attached to its periphery, which successively receives a jet from one or more nozzles. The jet on striking the cup is turned approximately 180 degrees in direction and leaves the runner with a low absolute velocity. Regulation of the wheel is accomplished by closing or deflecting the nozzle.

- **pelton wheel**
a type of impulse turbine, comprising a set of double cup-shaped buckets fitted on to the rim of a disc attached to a shaft, and operated by impact of a jet of water on the buckets from a nozzle. The flow of jet or jets being tangential to the wheel, it is also termed a tangential turbine.

- **pen**
a small space enclosed by any kind of fence which is used for confining pigs, cows and other animals.

- **pen plotter**
computer input device that allows a person to transfer an image into the computer by tracing or drawing.

- **pen type barn**
a modern barn designed for cattle in which the animals are not confined or stanchioned but have free run for exercise, loafing, and free access to roughage and water. The cows are moved a few at a time into a milking parlour arranged for the feeding of concentrates and for milking.

- **pendant**
a sheet metal disc bearing the observation point number and carried by the pendant wire.

- **pendant wire**
the wire exactly marking a section line and carrying pendants upon it to indicate the exact position of observation points.

- **pendular stage**
term suggested by Versluys who distinguished three stages in the zone of aeration. The pendular stage, the funicular stage and the capillary stage. In the first the water particles surround the points of contact of the solid particles, forming separate rings. In the second these water rings fit together and in the third they will fill the capillary channels.

- **pendular water**
if the soil particles are considered as touching spheres, the pendular water is that which is held in circles around the

point of contact of the spheres, but these are still air-spaces between the circles. At this stage, about 24 % of the pore space is occupied by water.

- **peneplain**
a relatively flat, featureless plain which has resulted from the erosion of former overlying formations first, by streams cutting a series of channels, and then the ridges which lay between such channels eroding until the resulting surface is almost flat.

- **penetrability**
the work required to push a probe a unit distance into the soil; method of measuring the physical characteristic of the soil.

- **penetrant**
wetting agents that enhance the ability of a liquid to enter into pores of a substrate, to penetrate the surface, are termed penetrating agents or penetrants.

- **penetration**
the hardness of a bitumen or asphalt determined by measuring the distance that a standard needle will penetrate vertically into a test sample of bitumen under specified conditions of temperature, loading and time.

- **penetration resistance**
1. number of blows of a hammer of specified weight falling a given distance required to produce a given penetration into soil of a pile, casing, or sampling tube.
2.unit load required to maintain constant rate of penetration into soil of a probe or instrument.
3.unit load required to produce a specified penetration into soil at a specified rate of a probe or instrument. For a proctor needle, the specified penetration is 2J in. and the rate is in per second.

- **penetrometer**
device for measuring the force required to push a probe rod into the soil; can be used for measuring the density or degree of soil compaction.

- **peninsula**
an elongated portion of land nearly surrounded by water and connected to a larger body of land, usually by a neck or an isthmus.

- **penny (nail)**
the term penny, when used to mark the length of nails, is supposed to be an abbre-

viation of pound. Thus, a four-penny nail was such that one thousand of them weighed four pounds, ten penny such that one thousand of them weight ten pounds, etc.

- **pentavalent element**
  any chemical element having five valence electrons.
- **perforated board (perf board)**
  non-conductive board with pre-drilled holes often used for assembly of permanent electronic circuits.
- **permanent-split capacitor motor**
  a type of capacitor-start motor where the capacitor and auxiliary windings remain in the running circuit while the motor operates.
- **phase angle**
  the angle describing the shift between the voltage and amperage in a circuit.
- **phloem**
  inner tree bark that transports plant food produced by photosynthesis to other parts of the tree.
- **photovoltaic effect**
  phenomena by which certain materials, such as cadmium sulphide, convert light energy into electrical energy.
- **physical property**
  characteristics of a metal that do not change during fabrication.
- **pictorial drawing**
  a type of drawing that shows an object as a unified whole.
- **piezoelectric effect**
  the phenomena by which certain quartz crystals produce electricity when subjected to pressure.
- **piping**
  the general term for pipes and/or hoses used in a hydraulic system.
- **piston**
  reciprocating primary engine component that compresses air or and air-fuel mixture in the combustion chamber and which receives the force from the expansion of these gases.
- **piston blow-by**
  the movement of combustion gasses around the piston and into the engine crankcase.
- **plough**
  a piece of primary tillage equipment that is used for breaking or turning over the soil.

- **ply**
  an individual layer in a sheet of plywood.
- **points of carbon**
  method used to indicate the carbon content of steel.
- **polarity**
  the condition of being positive or negative in accordance with some reference point.
- **pole**
  the moveable contact in an electric switch.
- **pole building**
  a type of building structure that uses wooden poles as the major structural support components.
- **port**
  a plumbing connection that allows flow into and/or out of a device.
- **Portland cement**
  a finely ground powder made from limestone, shale, clay, and iron materials that, when combined with water, forms the paste that bonds concrete together.
- **position**
  refers to the number of discrete positions the valving element in a hydraulic system directional control valve can assume in directing fluid flow.
- **positive displacement**
  a type of pump where the same volume of fluid is delivered per cycle, regardless of the output pressure.
- **potential energy**
  energy that is the result of a position.
- **potentiometer (pot)**
  a uneven resistor having three terminals.
- **power**
  the time rate of doing work.
- **power factor**
  1. cosine of the phase angle.
  2. ratio of true power to apparent power in an electrical circuit.
- **power-take-off**
  external tractor shaft for delivering rotary power to implements or other machines.
- **precision farming**
  information-based farming techniques where production inputs such as seed, fertiliser, and chemicals are applied only in affected areas and only in the amounts needed.
- **pressure override**
  in a hydraulic system, the difference between the cracking

and full-flow pressures for a pressure relief valve.

- **pressure tester**
  a diagnostic tool used to pressurise a liquid coolant system to test for leakage.
- **preventive maintenance**
  periodic maintenance practices intended to keep equipment in good working condition.
- **primary battery**
  batteries that are not designed to be recharged.
- **primary system**
  internal combustion engine system that is directly involved in creating compression and producing power.
- **prime mover**
  a device that, in a hydraulic system, powers the hydraulic pump.
- **printed circuit (PC) board**
  a non-conductive board with pre-drilled holes connected by copper traces. It is often used for assembly of permanent electronic circuits.
- **processing**
  transforming raw agricultural products into consumable food.
- **proton**
  the positively-charged portion of an atom.
- **protractor**
  a mechanical drafting tool for checking and establishing angles.
- **pruning**
  the process of removal of dead, broken, unwanted, diseased, or insect-infested wood.
- **PTO power**
  power available at a tractor's power-take-off shaft.
- **puddle**
  the molten pool formed when welding.
- **purlins**
  horizontal structural building supports that run perpendicular to poles or rafters.
- **pyrology**
  the scientific art of protecting forests from fire.
- **q**
  1. symbol for the flow, or flux of water.
  2. symbol for the rainfall temperature quotient.
  3. the disintegration energy or energy released per atom for a particular decay process or other nuclear reaction.

▪ **quadrant plate (machinery)**
plate which carries the stud wheels in the change wheel series.

▪ **quadratic response**
the output goes on increasing upto a certain level of input and then starts decreasing when input is increased further. In such cases the response is said to be quadratic and is given by:
$Y=a+bx+cx^2$.

▪ **quadruple cropping**
growing four crops in an agriculture year on the same piece of land, e.g., lady finger—cauliflower—potato and onion. Such a system is generally possible when adequate facilities for irrigation, fertiliser and other inputs are available and the produce can be marketed easily.

▪ **quagmire**
soft, wet, miry land, that shakes or yields under the foot.

▪ **qualitative**
relating to an object or feature in terms which, can not be expressed quantitatively, but has to be described in subjective terms.

▪ **qualitative analysis**
process to determine the nature of an unknown thing.

▪ **quality**
1. the excellence of the characteristic which the consumer desires in the product which he purchases.
2. a term of the approximate characterisation of radiation with regard to its penetrating power. It is usually expressed in terms of effective wavelength or half-life.

▪ **quality of irrigation water**
all irrigation waters contain dissolved salt. The salt content varies from 0.1 to 5.0 tons or more per acre-foot. A water quality test may alert the farmer to two hazards—the presence of salinity or the possibility of the formation of sodium soil.

▪ **quality of steam**
amount of moisture in the steam. Wet steam contains free moisture and is less than 100 per cent quality; dry saturated steam contains just enough heat to be all vapour and is 100 per cent quality; super heated steam has more than enough heat to change all its moisture into steam.

- **quality of waters**
quality of water means the suitability of water for irrigation use.

- **quantification**
the use of statistical methods in presenting, explaining and solving a problem in terms of quantifiable concepts and relationships, in handling data, and in producing objective system of classification.

- **quarry**
mine from where building stone, cement, rock, slate or similar materials are dug out.

- **quarry sap**
quantity of moisture contained in a freshly quarried stone.

- **quarry water**
water contained in the pores of rock when first quarried. It occurs mostly in stratified rocks and may be present in sufficient quantities to interfere with quarrying during freezing weather.

- **quarrying machine**
gang drill inclined at various angles for cutting channels, grooves, etc.

- **quartz (SiO2)**
silica or silicon dioxide, one of the major constituents of most common rocks. Has hardness of 7 and specific gravity of 2.65. Colourless, translucent or opaque, and forms double hexagonal pyramid crystals. Chief constituent of sandy fractions of soil.

- **quartz crystal**
silicon dioxide (SiO_2).

- **quartz lamp**
a mercury vapour lamp which uses in place of a glass container a tube of quartz.

- **quasi smooth flow**
a type of flow in rough conduits when the roughness elements are so close together that the flow essentially skims the crests of the elements.

- **quaternary period**
a geological period which includes both the Pleistoncene and recent periods comprising the second portion of the Cenozoic era, characterised by the rise of man and modern mammals.

- **quatrefoil**
a unit of decoration in the form of a four leaved flower.

- **quay**
a stretch of paved bank along a stream or other body of water, or other artificial land-

ing place constructed along, the bank, for the purpose of loading and unloading vessels.

- **quenching**
the process of dipping of heated steel into water, oil or other bath, to give necessary hardness.

- **quenching oils**
oils which are used in heat treating. Mineral, fish, vegetable and animal oils are usually compounded and sold under trade names.

- **quick**
a condition of the soil in which it is semiliquid, shifting and easily movable due to the presence of water.

- **quick break switch**
it is of the knife-blade type whose blade is made of two pieces. As the switch is pulled out the first half of the blade is withdrawn and as the throw increases, the second half is withdrawn by the action of a spring attached to the first blade.

- **quick condition (quicksand)**
condition in which water is flowing upward with sufficient velocity to reduce significantly the bearing capacity of the soil through a decrease in inter-granular pressure.

- **quick coupled portable laterals**
laterals used in sprinkler irrigation that are made up of lengths of pipe which are easily uncoupled to enable the line or lateral to be moved quickly to a new location.

- **quick coupler**
a connecting or disconnecting device for two lengths of pipes in a pipeline assemblage in a sprinkler method which allows quick assembly and separation of movable pipes.

- **quick growing**
pertaining to a plant in a particular climatic condition that matures in a comparatively short period.

- **quick opening valve**
a valve which has a revolving plug or a gate and a lever attached to it to facilitate sudden opening or closing.

- **quick return**
machines which are so constructed that the return stroke is much more rapid as compared to the forward or cutting stroke.

- **quick soil tests**
there are simple and rapid chemical tests of soils designed to give an approximation of the nutrients available to plants. Interpretation of results depends upon previous standardisation of soil tests with responses to fertilisers. It varies for different kinds of soil.

- **quick test**
a soil test in which the water content of the test specimen remains practically uncharged during the application of the confining pressure and the additional axial (or shearing) force.

- **quick water**
that part of a stream which has fall enough to create a decided current.

- **quicksand**
sand which has lost its grain-to-grain contact by buoyancy effect of water flowing upward through the voids. Such material, having some of the characteristics of a fluid possesses no load bearing value. It is a condition due to inherent instability of the sand not a type of material.

- **quill**
means a hollow shaft or spindle.

- **quill gear**
a gear or pinion cut on a quill or sleeve.

- **quintal**
a measure of weight; 100 kilograms.

- **quiscent zone**
the terminal part in the apical meristem of the root apex, where mitotic divisions are absent. May become meristematic if other part of meristem is injured.

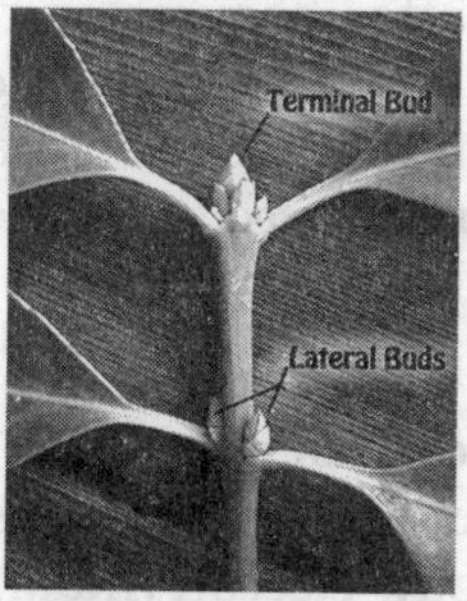

- **quonset**
a trade name for a building, now commonly applied to any one story round-roofed building when the roof is finished with metal roofing and extends nearly to the ground on both sides.

- **r**
rainfall factor in universal soil loss prediction equation.

- **r value**
the percentage of the crop land actually cropped in a year.

Frequencies of cropping in a fallow cycle (R) is equal to crop year times 100. 'R' values is synonymous to cropping index.

- **rabbet**

a rectangular groove or channel cut lengthwise in the surface of a plank to receive another member.

- **rabbing**

a process of burning of the residues of the crops in the field after harvesting of the crop. Rabbing is followed with an idea to keep the land free from weed seeds and pest etc.

- **rabi and monsoon sections of a canal**

these terms are used in Bombay-Deccan (India). The 'Rabi section' is the cross-sectional area of a canal in which assured supply is available for maturing two season and rabi crops only. The rest of the cross-section is the monsoon section in which water may be supplied when available, in the monsoon and rabi seasons without any assurance.

- **rabi crop (India and Pakistan)**

winter crop.

- **race**

a canal for conducting a current of water; a water course used to convey water for industrial purpose, as a mill race; it is called a HEAD RACE upstream from the point where the water is used, possibly for turning a water wheel, and a tail race downstream from the point of use; a raceway.

- **rack and pinion**

a device incorporating a toothed wheel and a toothed rod, to the bottom of which the swivel and current meter are attached, to provide rigid suspension and to insure the measurement of velocity at the desired point and depth.

- **rack losses**

grain lost on the straw rack of a combine by going through the rack or being carried over the end of the rack.

- **racking**

cleaning of the trash racks.

- **raddle rakes**

rakes used in the separating part of the hillside combine to move the straw.

- **radial drainage pattern**

drainage pattern resembling the spokes of a wheel in which

drainage channels flow radially either outward from a peak or inward toward a basin.

- **radial flow**
the direction of flow is along the radii of a tank from the centre to the periphery or vice versa. Also applies to a component of the flow.

- **radial flow pump**
a pump classified according to the direction of flow in which fluid enters the rotating impeller near the axis of rotation and discharges from the impeller into the casing with a combined radial and tangential motion. Fluid accumulating in the casing is passed continuously out the discharge nozzle. There is, in general, a continuous increase in pressure in the direction of motion.

- **radial flow sedimentation tank**
a circular sedimentation tank with the sewage or water entering at the centre through an upward flow pipe into a vertical walled drum.

- **radial flow turbine**
a type of turbine in which the path of any particle of water in its flow through the runner lies substantially in a plane perpendicular to the axis of rotation in a generally radial direction, the water being admitted at one radius and discharged at another radius usually all round the periphery.

- **radial force**
a force bearing in all directions from centre point.

- **radial gate**
a pivoted or hinged gate, the face of which is usually a circular are with centre of curvature at the pivot.

- **radial gate spillway**
overflow spillway utilising radial gates; when automatic, will automatically open or close with the rise or fall of the water surface in the canal.

- **radial honey extractor**
a centrifugal honey extracting machine in which the uncapped combs are arranged on a rotating reed like the spokes of a wheel with the bottoms of the combs toward the centre of rotation. The reel rotates inside a round tank having an outlet for honey at the bottom.

- **radial inward flow**
flow of water or other liquid in a circular tank, or water turbine, from the periphery to the

centre. The inlet or inlets are located at the periphery and the outlet or outlets at the centre.

- **radial outward flow**
flow of water or other liquid in a circular tank, or water turbine, from an inlet at the centre to an outflow at the periphery. The inlet or inlets are located at centre, and the outlet or outlets at the periphery.

- **radial piston pump**
a type of hydraulic piston pump in which the pumping chambers are perpendicular to the pump's drive shaft.

- **radial well**
a variation of strainer well; in this case the strainers are driven horizontally in the water bearing strata, radiating from a central sump or tube well. This enables tapping of a shallow depth of aquifer.

- **radiation**
heat transfer through electromagnetic waves.

- **radiation standards**
regulations that include exposure standards, permissible concentrations, and regulations for transportation.

- **radiator**
in the cooling system, the device that removes heat from water passing through it; it takes hot water from the engine and returns cooled water to the engine.

- **radioactive soil**
soil which possesses a minute degree of radioactivity. Radioactive phosphorus and other nutrient elements are sometimes added to the soil to study their functions in plant growth.

- **radioactive spring**
a spring whose water has an abnormally high and easily measurable radioactivity.

- **raft foundation**
a layer of concrete, usually reinforced, extending under the whole area of columns, piers and abutments and projecting outside their base periphery; used to provide a foundation fin case where the ground is unduly soft or the loading to be put upon it is unduly heavy.

- **rag iron**
the adjustable flat spring-like part of the corn sheller which

permits the shelling of both large and small ears. It serves to hold the ears against the shelling mechanism.

- **railing**
a method of brush removal from farms employing the use of railway rail.

- **rain**
drops of water in the air, generally larger than 0.02 inch in diameter, which fall to the earth at velocities usually greater than 10 ft per sec. in still air; drops of water that have become too large to be held in suspension.

- **rain forest**
an equatorial forest where the rainfall is in excess of 80 to 100 in.

- **rain belt**
a region of heavy or above average rainfall.

- **rain clouds**
clouds that contain enough nuclei to form ice crystals which may or may not be heavy enough to fall to the ground.

- **rain day**
used in some countries to denote a period of twenty-four hours during which more than a certain amount of rain is recorded. There are no universally accepted limits.

- **rain discharge**
the average rain (amount of in certain rain series) of the whole drainage area divided by time a seconds.

- **rain erosion**
erosion due to rain.

- **rain factor (rf)**
an expression of the humidity as a function of rainfall and temperature, as shown below:
RF= mean annual rainfall (mm) mean annual temperature (°C)

- **rain gage**
a device for catching and measuring the depth of rainfall. There are various kinds and sizes of rain gages, most of them employing the principle of catching the rainfall IDA collector of larger area than that of the measuring compartment, so that a given depth of water in the latter represents a considerably smaller depth of rainfall catch.

- **rain gun**
a nozzle that sprays water, sewage, liquid manure or industrial effluent into the air and out on to the land.

- **rain maker**
a rain simulator used for applying water in drops upon a field surface.

- **rain making**
methods of weather modification where artificial nuclei are supplied to the clouds to encourage precipitation.

- **rain pillars**
minor erosional forms composed of upward projecting pillars of soil or soft rock capped and protected, by pebbles or concretions.

- **rain season**
the period of the year in which monthly rainfall exceeds 10 per cent of the monthly potential evapotranspiration.

- **rain spell**
a rainy period of 15 consecutive days, each day recording at least 0.25 mm of rainfall.

- **rain storm**
storm associated with rainfall.

- **rain wash**
the creep of soil and superficial rocks under the influence of gravity and the lubricating action of rain.

- **rain water**
one of the purest of waters, apart from its gas content, which can be considerable. It is soft and contains carbon dioxide; therefore, it can dissolve lead and be poisonous if it passes through lead pipes or over a lead roof.

- **raindrop erosion**
a form of soil erosion resulting from soil splash caused by the impact of falling rain drops.

- **rainfall**
1. a fall of rain; precipitation in the form of liquid water.
2.the amount of rain, usually expressed in millimetres (inches) depth of water on an area, that reaches the surface of the earth.

- **rainfall area**
geographical extent of a given storm.

- **rainfall distribution coefficient**
the distribution coefficient for any storm is obtained by dividing the maximum rainfall at any point by the mean on the basin.

- **rainfall effectiveness**
the ability of rainfall to promote vegetation; governed by the amount of rain absorbed by the soil and thus made available to the plant roots.

- **rainfall excess**
that part of a rain of a given storm which falls at intensities exceeding the infiltration capacity and is thus available for direct runoff.

- **rainfall index**
the average rainfall intensity above which the volume of rainfall equals the volume of observed runoff.

- **rainfall infiltration**
the portion of rainfall which passes, downward from the surface into the soil or rock. It is equal to the total precipitation less the losses due to interception by vegetation, depression or other storage, evaporation from all moist surfaces, and surface runoff.

- **rainfall intensity**
the rate at which rainfall occurs expressed in depth units per unit time. It is the ratio of the total amount of rain to the length of the period in which the rain falls.

- **rainfall intensity curve**
a curve which expresses the relation of rates of rainfall and their duration. Each curve is generally for a period of years during which time the intensities shown will not, on the average, be exceeded more than once.

- **rainfall intensity frequency or rainfall recurrence interval**
the average time interval between the occurrence of rainfall of a given intensity and that of an equal or greater intensity.

- **rainfall interception**
interception of the fall of raindrops by a canopy of vegetation or vegetative residue which is usually expressed as ins. depth retained.

- **rainfall penetration**
the depth below the soil or rock surface reached by a given quantity of rainfall which has passed below such surface. The maximum extent of penetration is down to ground water.

- **rainfall province**
an area characterised by similarity in distribution and seasonal concentration of precipitation.

- **rainfall rate**
the amount of precipitation occurring in a unit of time, generally expressed in inches per hour.

- **rainfall simulator**
it is a device which can reproduce, accurately and repeatedly, artificial rain which has precisely the same effect on the soil as natural rain.

- **rainfed farming**
growing of field crops entirely with rain water received during the crop season (rainfall usually 800 m) under humid and subhumid climate and the crops may face little or no moisture stress during their life cycle.

- **rainfed stream**
a stream the water in which comes primarily from surface runoff resulting from rainfall.

- **rainout**
removal of particles or grass from the atmosphere during the formation of raindrops on nuclei. Rainout takes place in the cloud or near it; washout takes place further down, nearer the earth.

- **rainsplash (raindrop erosion)**
soil erosion caused by falling raindrops. A series of impacts from the raindrops dislodges the soil particles and bounces or washes them down slope. If the process continues the impact of the raindrops compacts the soil and increases the surface flow of water; this reduces the water that actually percolates down into the soil.

- **rainy**
designating a period in which rain falls more or less continually.

- **rainy season**
the season of heavy or more than normal rainfall as contrasted to the dry season of little or no rainfall.

- **rainy spell**
frequent rainfall for a protracted period of time, such as for several days.

- **rainy weather**
a period of relatively short duration during which there are numerous showers of rain.

- **raised beach**
a shelf or terrace of shingle, gravel, and sand elevated above the present level of the lake or sea in which it was formed, and indicating a change in the relative level of land sand water surface.

- **raised bed sunken bed technology**
  it involves construction of 20 cm high and 3 m wide raised beds for upland crops adjacent to 20 cm deep and 6 m wide sunken beds for paddy crop along with a tank for the collection of overflow of water to provide *in situ* drainage and storage of water. It has been developed for the management of fine textured slowly permeable flat valley soils occurring in the agro climatic regions receiving more than 1000 mm rainfall during rainy season.
- **raised moss**
  a class of peaty land consisting of sphagnum which rises from the edges towards the middle and thus shows a convex upper surface.
- **ram**
  a single-acting hydraulic cylinder with a rod diameter nearly equal to the cylinder bore.
- **reaper**
  an early piece of farm equipment used to cut and gather grain crops.
- **rectification**
  the process of conversion of alternating current to direct current.
- **refining**
  the removal of unwanted impurities.
- **refrigeration**
  keeping cool.
- **remote sensing**
  gathering and recording data from a distance.
- **repulsion-start induction-run motor**
  a type of wound-rotor AC electric motor capable of starting very heavy loads.
- **reservoir**
  hydraulic system component for storing and cooling hydraulic fluid.
- **resistance**
  characteristic of any material that opposes the flow of electricity. It is measured in ohms.
- **resistor**
  an electronic or electrical component that is intended to introduce resistance into a circuit.
- **reverse-biased**
  the condition where the cathode of an electronic device is more positive than the anode.
- **rheostat**
  a variable resistor having two terminals.

- **ridge cut**
a cut made at the top end of a rafter. It is also called the upper plumb cut.

- **rigid arch building**
a type of building in which a series of wall and roof arch units form the building structure.

- **rise**
the height of a roof.

- **roof slope**
the rise per foot of run for a roof.

- **root cutting**
section of a root cut and used for propagation purposes.

- **root mean square (rms)**
the effective voltage of AC electricity that is equal to 0707 times the peak voltage. It is the type of voltage measured by most AC voltmeters.

- **root pruning**
systematic cutting of the roots by hand or machine so as to encourage the roots to develop within the root ball range.

- **rotary actuator**
a hydraulic device that produces rotating motion, such as a hydraulic motor.

- **rotor**
the rotating component of an AC electric motor.

- **run**
the horizontal length of the roof slope.

- **running winding**
the main winding on the stator of an AC electric motor.

- **s**
1. south pole of magnet.
2. length of stroke of engine piston (usually in inches).
3. degree of slope factor in universal soil loss prediction equation (USLE).

- **s matrix**
the S matrix of a soil material is the material within the simplest peds, or composing apedal soil materials, in which the pedological features occur; it consists of the plasma, skeleton grains and voids that do not occur in pedological features other than plasma separations.

- **s wrench**
an S shaped wrench having either fixed or adjustable openings.

- **sa**
symbol for an accumulation of soluble salts other than calcium sulphate or calcium car-

bonate. Used as a subscript to the designation of any horizon and in its manner of use is comparable to that described for ca or cs. If the symbol is used, the horizon must have more salt than the parent material is presumed to have had.

- **sabbing cutter**
a wide faced, milling cutter having nicked teeth to allow an easier cut than is possible with a plain toothed cutter.

- **sacking spout**
a spout through which material will pass in order to fill sacks. Such materials may be flour or grain, from a threshing machine, combine, mill or elevator. The sacks are firmly attached around the periphery of the spout.

- **saddle**
a steel or concrete structure, used for supporting a pipe or penstock laid above the surface of the ground.

- **saddle brack**
a hill or ridge having a concave outline at the top.

- **saddle bracket**
bracket on the top of a telegraph pole to support the insulator.

- **saddle duster**
a manually operated insecticide duster which is placed on a horse or mule so that the operator may ride and dust a row on each side; also called saddle gun.

- **saddle flange (pipe fitting)**
curved flange hollowed out to fit a boiler.

- **saddle key (mach)**
key whose inner face is hollowed to fit its shaft.

- **saddle siphon or hood siphon**
a siphon having its dust formed of an U-shaped hood. The inlet lip of the hood is submerged below the full reservoir level and that of the outlet lip terminates in a water seal formed by various devices. The siphons are, sometimes, named according to the water seal provided.

- **safe bearing capacity**
the load usually expressed in ton & per square foot, which a given soil or foundation can safely support without appreciable settlement or movement.

- **safe edge**
an uncut edge on a file; an edge presenting a smooth surface,

without teeth, so that the file may be used in a corner and cut on one surface only.

- **safe velocity**
it is the highest velocity at which water may move in a channel or through a conduit without causing erosion or scour. Same as permissible velocity or non-eroding velocity.

- **safe working boiler pressure**
the highest pressure to carry on a steam boiler.

- **safe yield**
with reference to either surface water or groundwater supply, the rate of diversion or extraction for consumptive use which can be maintained indefinitely, within the limits of economic feasibility, under specified conditions of water supply development.

- **safe yield of reservoir basin**
the annual runoff flowing from the basin into the reservoir that can be depended upon for utilisation.

- **safety catch (elevator construction)**
device which works by powerful springs, to relieve stops on the bottom of an elevator car which grasp the guides.

- **safety factor**
the ratio between the ultimate load and the design load for materials, structures or mechanisms. This provides a measure of safety to cope with load in excess of design load without failure.

- **safety gate**
a gate installed in a high-level navigation canal for closure in emergency to prevent loss of water.

- **safety hoist**
in machinery, any hoist constructed with differential pulleys; also an ordinary rope hoist in which on automatic catch or safety stop secures the apparatus against running down.

- **safety release hitch**
an attachment which is designed to release a pulled implement, such as a plough, mower, etc., from a tractor when the implement strikes an obstruction.

- **safety shield**
shield used to protect a person from rotating shaft on machinery.

- **safety switch**
a switch, usually of 30 amperes or more capacity, enclosed in a grounded metal box, and so arranged that the door of the box cannot be opened unless the switch is off.

- **sag correction**
the amount by which soundings must be reduced to obtain true vertical values.

- **sag rod**
a number used in tension as for sustaining a suspended weight.

- **sagebush**
a bush growing in arid regions, olive green in colour; growing generally in strong rich soil, it does not thrive on poor, or strongly alkaline soil, nor in extremely dry desert regions.

- **sagger**
a fire-clay box wherein delicate pieces are placed while being baked.

- **sajong or shachiang soils**
soils containing 85 per cent calcium carbonate and having lime concentrations of ginger root size.

- **salic horizon**
a horizon at least 6 inches thick with secondary enrichment of salts more soluble in cold water than gypsum.

- **salina**
a salt march, or salt pond, enclosed from the sea.

- **saline**
something containing salt.

- **saline alkali soil**
a soil which contains both sufficient exchangeable sodium to inhibit the growth of most crop plants and appreciable quantities of soluble salts. The exchangeable sodium percentage is greater than 15, the conductivity of the saturation extract is greater than 4 milliohms per centimetre at 25°C, and the pH of the saturated soil is usually less than 8.5.

- **saline alkaline water**
water which has moderate concentration of soluble sodium salts.

- **saline contamination**
contamination by intrusion of salt water.

- **saline intrusion**
movement of salt water underground from the sea coast inland. Since salt water is denser than fresh, it passes below the fresh water, forming a 'saline wedge'. Excessive

pumping of deep wells may extend the wedge inland to produce saline well water.

- **saline salty**
nitrates, chlorides and phosphates are the commonest salts in sewage that may have to have their concentrations reduced by advanced waste water treatment or de-nitrification.

- **saline sodic soil**
a soil containing sufficient exchangeable sodium to interfere with the growth of most crop plants and containing appreciable quantities of soluble salts. The exchangeable sodium percentage is greater than 15 (or SAR greater than 13), the conductivity of saturation extract greater than 4 dS/m (25°C), and the pH is usually 8.5 or less in the saturated soil.

- **saline soil**
a non-alkali soil, which contains soluble salts in such quantities that they inhibit the growth of most crop plants. The conductivity of the saturation extract is greater than 4 milliohms per centimetre at 25°C, the exchangeable sodium percentage is less than 15, and the pH of the saturated soil is usually less than 8.5.

- **saline soil reclamation**
process of reclaiming saline soil by reducing its salt content; consists of application of sufficient quantities of water to the soil to dissolve the salts and leach them out in drainage; salts washed out by flooding the soil with river water and the salt content reduced.

- **saline spring**
a spring whose water contains a considerable quantity of sodium chloride.

- **saline water**
water which contains moderate concentration of total dissolved salts.

- **salinisation**
the accumulation of soluble salts at the surface or at some point below the surface of the soil profile.

- **salinity**
the relative concentration of salts, usually sodium chloride, in a given water. It is usually expressed in terms of the number of parts per million of chlorine (Q).

- **salinity control**
abatement or prevention of salt water contamination of agricultural, industrial and municipal water supplies, or

reducing alkaline salts and preventing deterioration of cultivable lands.

- **salinometer**
an instrument for measuring the amount of salt in a solution; a hydrometer especially graduated so as to indicate directly the percentage of a particular salt (especially common salt) in a solution of it; the process is called SALINOMETRY.

- **salminen probe**
a probe for taking soil samples. It is of a stem one meter long, pointed with a handle, with a cavity 30 cm long in the lower part; one of the edge of the cavity is sharpened and slightly protuberant; it is sunk into the soil and turned, then removed; the soil sample lies inside.

- **salometer**
a device, usually a hydrometer, with special graduations in degrees or percentages used for determining the strength or salinity of a particular salt solution. Four degrees on the salometer equal one per cent of salt in the solution.

- **salt**
salts are formed when an acid reacts with an alkali. They form ions in water. Organic salts are called ESTERS. Table salt is sodium chloride, NaCl.

- **salt affected soil**
a soil which has been adversely modified for the growth of most crop plants by the presence of soluble salts, exchangeable sodium, or both.

- **salt alum tanning**
a simple method of tanning sheep skins whereby the previously salted and fleshed skins are treated overnight with a 2 to 1 mixture of salt and alum and later on with Neat's foot or glycerine.

- **salt balance**
difference between total dissolved solids brought to land annually by irrigation water and total solids carried away annually by drainage water.

- **salt encrustation**
overlaying by salt.

- **salt glaze tile**
a tile that has been glazed by throwing salt in the kiln during the baking process.

- **salt index**
it is used for determining the suitability of irrigation waters containing salts in excess of 6CO ppm, assuming that

plants can tolerate a maximum of 1170 ppm concentration of sodium chloride.

SI=(total Na-24.5)—Total Ca-Ca in $CaCO_a$ & $Ca(HCO_3)_2$ x 4.85

In this index all quantities are expressed in parts per million, and all values of magnesium are taken equivalent to calcium. The salt index (SI) is negative for all good quality waters and positive for all waters unsuitable for irrigation. Several objections have been put forward against wide applicability of salt index in judging the quality of irrigation water. It is not necessary that all the good waters will have values between —24.5 and zero because the negativity of SI will improve with soluble calcium. Moreover, as the effect of magnesium on soil properties is not the same as that of calcium, assuming Mg equivalent to Ca does not appear to be sound enough.

- **salt marsh**
  a marsh periodically flooded by salt water.

- **salt method**
  a method of measuring the discharge of water which flows through closed passage under pressure. A salt solution of known strength is injected at a known rate into the water at one point, and samples of the water are taken at a lower point and analysed.

- **salt respiration**
  the increased respiration over the normal or ground state respiration following transfer of a plant from water to a salt solution.

- **salt stress**
  this stress results from an excess of any one of a large number of salts.

- **salt tolerance**
  the ability to tolerate high concentration of soluble salts in the root medium. It usually is expressed by reference to the salinity level that causes a certain decline in yield.

- **salt velocity method**
  the measurement of discharge by injecting salt at one point and observing the time required for the salt solution to reach another point.

- **salt ventilation**
  systematic discharge of salt solution, formed in self-mineralisation of irrigation waters circulating in the soil.

- **salt water underrun**
  the more rapid movement of the bottom water in tidal estuaries, making the bottom water saltier than the top water.
- **saltation**
  stream transportation of elastic sediments by intermittent leaps or bounds.
- **saltation layer**
  the zone above the bed within which phenomenon of saltation occurs in flowing channel.
- **saltation load**
  the material bouncing along the bed, or moved, directly or indirectly, by the impact of the bouncing particles.
- **saltation load discharge**
  the amount of material moved in unit time due to saltation.
- **salted soil**
  a generic term applying to all soils having characteristics which are caused by exposure to excessive amounts of soluble salts.
- **salt-water encroachment**
  the gradual entrance of salt water into fresh water aquifers.
- **salt-water intrusion**
  the phenomenon occurring when a body of salt water, because of its greater density, invades a body of fresh water. It can occur either in surface or ground water bodies. The balance between the two is expressed by the Ghyben-Herzberg formula.
- **salvage**
  the utilisation of waste materials.
- **sample**
  a small part of a large quantity, one that truly represents the whole.
- **sample mean**
  the arithmetic mean value of a variate in a sample or in the plots of a particular experiment; when there is no danger of confusion, called simply the MEAN.
- **sample strip**
  long narrow strip employed as a sampling unit in surveys.
- **sampler**
  any of various devices which is used for taking a representative small portion, as from a bag, a ear of grain, a can of milk, a tub of butter, a cheese, etc.
- **sampling**
  the procedure of selection of a sample from population is known as sampling.

- **sampling block**
  selecting a sample of the population by first grouping the units of investigation into contiguous blocks, the numbering the blocks and drawing a sample of blocks. Within the blocks chosen, the entire population may be included in the sample.

- **sampling distribution**
  the distribution of a statistic (or set of statistics) in all possible samples which can be chosen according to a specified sampling scheme.

- **sampling plan**
  the procedure for acceptance on the basis of sampling information, e.g., single-sampling plan, double-sampling plan, multiple-sampling plan and sequential-sampling plan.

- **sampling purposive**
  selection of a portion of the population which has the same average or other measures as the universe in regard to one or more characteristics which are already a matter of statistical knowledge.

- **sampling tube**
  any of several tube-like devices, commonly made of brass or stainless steel, of open or closable types which is used for extracting representative samples of grains, feeds, seeds, liquids, gases, etc.

- **sampling unit**
  the ultimate unit of assessment of measurement in a sample. It may be the individuals, a given area of ground, etc.

- **sampling, proportional**
  selecting a sample in which all significant elements are represented in the same proportion as they occur in the universe.

- **sand**
  a group of textural classes in which the particles are finer than gravel but coarser than silt. Such soils provides larger spaces between soil particles and permits a freer movement of water and air throughout the soil mass.

- **sand and boll screen**
  a device for cleaning cotton seed which consists of an inclined reel covered with revolving screens of various sized meshes to separate seeds and foreign matter according to size.

- **sand bank**
  a covering of sand around and above structures built in the diversion channel to absorb most of the shock from the

blast of the solid bulkheads and to prevent destructive cracking or displacement of the structure. The method using such material is known as sand bank method.

- **sand binder**
any plant which can hold sand in its place and prevent it from shifting by the grip of its roots.

- **sand blasting**
a method of removing all surface impurities from old metal surface and consists of forcible driving under pressure of hard particles of sand, causing controlled erosion of the surface against which the blast is directed. The treatment not only removes all surface impurities, including tight mill scale, but also imparts an etched texture and a largely increased bonding area. If steel grit (crushed shot) is used instead of sand, the method is called GRITBLASTING.

- **sand clay**
soil which contains more than 45 per cent sand, more than 35 per cent clay, and less than 20 per cent silt.

- **sand clay (road surface)**
a surface composed of a mixture of sand and clay where the two materials have been blended, so that their opposite qualities tend to maintain a condition of stability under varying moisture content. Some deposits are found in nature, but the bulk of such surfaces are prepared artificially.

- **sand gate**
a sluice gate used for sluicing out sand and sediment collected at a sand trap in an irrigation canal.

- **sand island method**
a patented construction method used in streams for piers; etc., by forming a special island of sand in the stream to provide a stable footing through which a pier may be sunk to rock or other sound formation.

- **sand pump**
a long cylindrical tube fitted with a valve at the bottom and plunger with a valve or valve-action leathers, and open at the top, used for raising up the mud which accumulates at the bottom of a boring during the sinking process.

- **sand river**
a dry river bed beneath which there is water, common in the wadis of African and Arabian deserts, which enjoy running

water only occasionally, after heavy rain.

- **sand spit**
  a narrow sand embankment, created by an excess of deposition at its seaward terminus, with its distal end (the end away from the point of origin) terminating in open water.

- **sand streaking**
  a surface defect in which fine materials form a thin layer. This is due to lean, harsh, wet mixes with bleeding tendencies or poorly graded sand deficient in fines, coarsely ground cement, and leaky forms.

- **sand wash filter**
  a moving bed sand filter.

- **sandbag**
  a bag filled with sand or earth used to form a temporary dam. Sometimes filled with concrete and used for the same purpose.

- **sandboil**
  1. a spring or geyser in a leveed area due to pressure of flood water around the outlet.
  2. a spring or geyser in an excavation caused by unbalanced hydrostatic pressure.

- **sanding**
  the process of filling up or raising the bed of a body of water through the accumulation of sand; accretion of sand.

- **sandstone**
  rocks of sedimentary origin and which have resulted from the consolidation of sand and gravel beds which either contained the cementing substance or were impregnated with it during the process of solidification; for the most part, consist of quartz grains cemented together by silica, clay, iron oxide, or lime carbonate.

- **sandstorm**
  windstorm which carried large amounts of sand and reduce visibility to a great extent.

- **sandwich brick, the lining**
  lining consisting of a layer of plaster on the subgrade and two layers of tiles with an impervious layer of cement mortar sandwiched in between.

- **sandy clay loam**
  this class of soil contains 20-35 per cent clay, less than 28

per cent silt and 45 per cent or more of sand.

- **sandy loams**
soil material that contains either 20 per cent clay or less, and the percentage of silt plus twice the percentage of clay exceeds 30 and 52 per cent or more sand, or less than 7 per cent clay, less than 50 per cent silt, and between 43 per cent and 52 per cent sand.

- **sandy soil**
a broad term for soils of the sand and loamy sand classes where more than 70 per cent sand and less than 15 per cent clay are present.

- **sanitary**
means pertaining to, or tending to promote health.

- **sanitary fitting**
pipe fitting used on the milk piping systems of dairies. These fittings are designed so they can easily be disassembled for cleaning.

- **sanitary pump**
any of several kinds of pumps for the handling of milk and other liquids whose parts can be easily removed for thorough cleaning and sterilisation and which is made of a metal which does not affect the product being handled.

- **sanitary trap**
a part of a vacuum pump installation in a milking machine which is designed for collecting and holding unwanted materials. It can be easily emptied and cleaned.

- **sanitation**
means biological cleanliness such as clean air, clean water and clean food for both man and beast.

- **saprobic classification**
the Kolkwitz-Marsson saprobic system, a pollution classification for rivers, dividing the recovery zone into oligosaprobic, mesosaprobic and polysaprobic, with subdivisions.

- **saprobicity**
the effects of organic matter on life in rivers gaprobiotic. Feeding on dead or decaying organic matter.

- **saprophyte**
a saprobe that feeds on soluble organic matter. Many bacteria and fungi are saprophytes.

- **saprozoic**
descriptive of saprobes, which feed on solid organic matter, like most animals.

- **sapwood**
  the softwood that is located between the inner bark of a tree and the heart-wood. It is also called xylem.

- **SAR**
  sodium absorption ratio. Ions are expressed as milliequivalents per litre. Good irrigation water has an SAR less than 4; SAR of 7-8 leads to alkalinity and at SAR 16-18 there is high hazard from use of the water.

- **sarda type fall**
  an unflumed, vertical drop structure, suitable for drop not more than 6.0 feet, evolved on the Sarda Canal system (India) having raised rectangular or trapezoidal crest.

- **sash house**
  a frame structure covered with window sash to provide an inexpensive hotbed or small greenhouse. It is commonly used to start plants for later transplanting.

- **satellite treatment**
  the arrangement of small sewage plants around a city to treat mainly domestic sewage, possibly with the return of the sludge through a sewer to a central treatment plant.

- **saturate**
  1. to fill all voids between soil particles with liquid.
  2.to form the most concentrated solution possible under a given set of physical conditions in the presence of an excess of the substance.

- **saturated air**
  air containing all the water vapour that it is capable of holding at a given temperature and pressure.

- **saturated ammonia vapour**
  vapour of ammonia (NH_4) which contains an amount of heat energy to exactly vaporise every particle; there is no free liquid and no superheat.

- **saturated flow**
movement of water through soil that is saturated with water as in irrigation or during a rain storm; mostly gravitational water.

- **saturated hydrocarbon fuel**
fuels, such as gasoline and diesel, that contain the maximum possible number of hydrogen atoms and that contain only single covalent bonds.

- **saturated liquid**
a liquid which contains at a given temperature as much of a solute as it can retain in the presence of an excess of that solute.

- **saturated rock**
a rock which has all its interstices or void spaces filled with water.

- **saturated salt solution**
a salt solution containing all the salt it can hold in solution at a given temperature. Addition of salt to the solution will result in precipitation of salt.

- **saturated soil**
soil holding the amount of base in order to have a selected pH; such a soil is said to be saturated up to that pH.

- **saturated soil paste**
a particular mixture of soil and water. At saturation, the soil paste, as it reflects light, flows slightly when the container is tipped, and the paste slides freely and cleanly from a spatula for all soils at its saturation percentage.

- **saturated solution**
a solution containing maximum amount of solute dissolved in it at a particular temperature and pressure. A state of super saturation can be created, but such solutions are unstable and may precipitate.

- **saturated steam**
all of the water in the form of vapour and at a temperature such that lowering of the temperature would cause a condensation of part of the steam and raising it would cause superheat.

- **saturated unit weight**
the wet unit weight of a soil mass when saturated.

- **saturation**
1. completely filled; a condition reached by a material, whether it be in solid, gaseous, or liquid state, which holds another material within itself in a given state in an amount such that no more of such

material can be held within it in the same state. The material is then said to be saturated or in a condition of saturation.

2. the condition of a liquid when it has taken into solution the maximum possible quantity of a given substance at a given temperature and pressure.

- **saturation capillary head**
the distance from the free water surface to the highest elevation at which complete saturation exists.

- **saturation curve**
the curve showing the zero air voids unit weight as a function of water content.

- **saturation deficit**
the difference between the quantity of a substance in solution and the quantity of such substance required to create a saturated condition, at a given temperature and pressure, usually expressed in terms of percentage of the difference to that required for saturation.

- **saturation extract**
the solution extracted from a soil at its saturation percentage.

- **saturation humidity**
relating humidity of 100 per cent which also refers to saturation vapour pressure.

- **saturation index**
an estimate of carbonate precipitation from irrigation water as a function of the degree of calcium carbonate saturation of the : soil solution.

- **saturation line**
line of saturation, percolation line or hydraulic grade line. Line across the banks on either side of a canal in filling or partly in filling and partly in cutting up to which the banks get saturated after the canal has been running for some time.

- **saturation percentage**
the moisture percentage of a saturated soil paste, expressed on a dry-weight basis.

- **saturation point**
the point beyond which no further additions can be held, utilised, or accommodated.

- **saturation pressure**
see saturation vapor pressure.

- **saturation vapour pressure**
The vapour pressure of a system, at a given temperature, wherein the vapor of a sub-

stance is in equilibrium with a plane surface of the pure liquid or solid phase of that substance. The term refers to the vapor pressure of a system that has attained saturation but not subsaturation. It is usually measured in relation with water.

- **saturation zone**

that portion of the lithosphere in which the functional interstices of permeable rock or earth are completely filled with water under positive hydrostatic pressure, that is pressure in excess of atmospheric pressure. It is the portion of the lithosphere that supplies water to springs and wells and which contains the groundwater. Its upper surface is the piezometric surface as measured in observation wells.

- **saturator**

a pressure chamber in which air can be injected into water at (say) 4 atm pressure, e.g., as a part of dissolved air flotation. A saturator can be either unpacked (empty) or packed with stones or plastics medium. Packing helps to break up the bubbles and flow pattern. The detention period is normally from 30 to 60s.

- **saucisse**

a bundle of brushwood of stouter and longer material than in a fascine but used for the same purposes.

- **sausage dam**

a type of deflection or diversion dam composed of layers of boulders, overlaid and held in place by hogwire fencing; distinguished from a basket dam, in that the latter is a woven wire sack filled with boulders and piled up as in a sack dam. Both are especially adapted to jetties and riverbank protection.

- **sausages**

one form of stone mesh, consisting of tubes of wire mesh, usually circular or elliptical in section, filled with stone and closed at the ends. In one form, somewhat similar to bolsters used for protection of river banks, when the sausages are laid parallel and touching down the sloped bank and out into the bed, or horizontally as a mattress or apron or to form a chute. May be laid over a layer of gravel, which acts as an inverted filter. Form a flexible and rapidly constructed protection. Trees or shrubs may be planted among the sausages to bind the stone with their roots.

- **savannah**
a tropical open, level grassland of tall grasses, interspersed with trees and shrubs. The vegetation is drought tolerable; able to withstand a hot dry season of considerable length.

- **savannah climate**
a zone which is warm throughout the year with maximum temperature occurring during the rainy season. The temperature conditions can accommodate a year round growing season for most tropical and mesothermal crops.

- **saw**
1. a strip of steel having sharp tooth projections, which when drawn back and forth is used to cut wood or metal. It is also made in circular disk form and in endless band form.
2.teeth which project into and form the bottom of the ginning roll box of an air blast cotton gin.

- **saw dust**
it can be used as a mulch, bedding material, compost, or for direct application to the soil. As a fertiliser material, sawdust ranks very low. Sawdust contains 4 kg of N, 2 kg of P_20_5 and 4 kg of K_ZO per ton of material, on an oven-dry basis. Alfalfa is from 5-15 times as plentifully supplied with the essential elements, as is sawdust.

- **saw set**
an instrument used to set or turn the teeth of a saw a little outward, that they may make a kerf somewhat wider than the thickness of the blade, to prevent clogging or friction.

- **saw timber**
standard trees large enough to be economically cut, and made into lumber.

- **sawbuck**
1. a mountain range having sharp peaks of about equal height; a sierra.
2.one or two X-shaped frame sometimes called a saw horse used for supporting poles and small log to be sawed.

- **sawdust**
dust or small particles or shavings of wood, etc., obtained from sawing.

▪ **sawdust concrete**
a type of nailing concrete produced by Portland cement, sand and sawdust.

▪ **saybolt viscosity**
the time in sec. it takes 60 c.c. of oil to flow through a Saybolt Universal Viscosimeter under specified conditions.

▪ **scab**
a short board or plank fastened to the side of the butt joint of two structural members. It keeps the members in line and gives strength to the joint.

▪ **scab lands**
areas where the soils cover is shallow and the rock lies close to the ground surface, is parallel to it, and frequently even exposed. Soils on scab land are so shallow as to cause serious runoff and erosion problem.

▪ **scaffold**
a temporary erection used in the construction, alteration or demolition of a structure, to support or to allow of the hoisting and lowering of workmen, their tools and materials.

▪ **scalding vat**
a tank of very hot water for dipping of hogs, poultry, vegetables or fruits, causing hair, feathers or peels to loosen for easy removal.

▪ **scale**
a drafting term indicating the relationship between the dimensions of the drawing and of the drawn object.

▪ **scale effect**
a conventional expression for possible discrepancies involved in extrapolating the results obtained on a model to its prototype.

▪ **scale factor**
the number represented from the origin on the x-axis equal to the length of one cycle on the y-axis for a semi logarithmic plot.

▪ **scale ratio**
the ratio of a dimension on a model to the corresponding dimension of its prototype. It may represent the ratio of lengths, velocities, discharges, forces or such other dimension as may be specified.

▪ **scale welding**
a special purpose technique used in making fillet welds in which distortion or over heating of a thinner section becomes a problem. A puddle is created, some welding rod added and the puddle enlarged,

and then the flame is withdrawn and the puddle allowed to solidify. A new puddle is created overlapping the first and the operation is repeated.

- **scale, fur**
the whitish crust formed inside kettles or hot water systems when the carbonate hardness of water is deposited by boiling. It is mainly calcium carbonate, $CaCO_3$.

- **scaling**
prying loose pieces of rock off a face or roof to avoid danger of their falling unexpectedly.

- **scalloped edged rolling coulter**
a rolling-type coulter with notched or scalloped edges which is designed for the cutting of heavy trash.

- **scalper**
a machine or device used in flour or feed mills for the removal of the outer seed covering of grains.

- **scalping**
removal of large lumps at an early stage of mineral dressing. Sometimes in a mechanical sorting plant it may mean the early removal by magnetic separators of iron and steel pieces larger than about 20 cm. removing the larger particles of dust in an expansion chamber or other inexpensive device before passing the dust-laden gas to the units that remove fine dust, electrostatic precipitators, fabric filters, high-energy scrubbers, etc.

- **scalping screen**
the top sieve on a cleaning or sorting machine. Its purpose is to take off large foreign objects from the material being cleaned or sorted so the lower screens can do a better job.

- **scantling**
small timber as 2 by 3, 2 by 4, etc., used for studding.

- **scarf weld**
a joint that is made by overlapping and welding together the scarfed ends or edges of metal rods, sheets, etc.

- **scarified soil**
soil loosened to a shallow depth.

- **scarify**
in soil tillage, it pertains to loosening the top particles by means of machines equipped with a set of steel teeth. Designed to penetrate the soil when moved across the surface.

scarifying
1. a process for scratching or scarring hard seeds of crops like legumes to facilitate absorption of water and to speed up germination. This process is usually taken up after hulling.
2. scratching, loosening to a shallow depth or crushing hard uneven surface of soil to provide for more tillable soil.

scarp
1. a line of cliffs produced by faulting or erosion.
2. a relatively straight, cliff-like face or slope of considerable linear extent that breaks the general continuity of the land by separating surfaces lying at different levels.

scarp beach
an almost perpendicular slope along the beach foreshore, an erosional feature due to wave action. It may vary in height from a few inches to several feet, depending upon wave action and the nature and composition of the beach.

scatter diagram
it is more particularly used to show the cluster or scatter at various ranges of the correlation between the variates.

scattering disks
round metal plates with several bars across the top of the plates mounted in a horizontal position beneath the hopper on a seeder or end-gate lime spreader to distribute and spread the lime or seed as it falls from the hopper. This is done by rotating the scattering disks so that when material strikes them it is propelled outward.

scavenge
to remove burned gases from the cylinder of an internal combustion engine after a working or power stroke.

schematic drawing
a type of graphic communication that uses standardised symbols, or schematics, to represent objects.

schematic symbol
a standardised symbol that represents an object.

science
systemised knowledge of nature and the physical world.

s-curve or s-curve hydrograph
a graph showing the summation of the ordinates of a series of unit hydrographs spaced at unit-rainfall duration intervals. It represents the hydrograph of unit rate of

rainfall excess continued indefinitely. If the ordinates of the curves are expressed as a percentage of the total unit hydrograph volume and the abscissa in unit of time or in percentage of lag time, the S-curve is known as a summation graph.

- **sealer**
  an electronic device which produces an output pulse for a given number of input pulses. Term generally applied to a device for indicating the total number of observed pulses or events.

- **secondary battery**
  a rechargeable battery.

- **semiconductor**
  a material that is neither a good insulator or conductor.

- **series circuit**
  an electrical circuit containing two or more loads and having only one path for the flow of current.

- **series-parallel circuit**
  an electrical circuit containing both series and parallel electrical loads.

- **service conductors**
  electrical conductors from the service point to the service disconnecting means.

- **service drop conductors**
  overhead electrical conductors originating at the utility company service transformer.

- **service entrance equipment**
  the components of the electrical system needed to carry current form the building service conductors to the point of origin of branch and feeder circuits within the building.

- **service entrance panel**
  the main control and cut-off means for the electrical service in a specific building or load centre.

- **service factor (SF)**
  indicates the continuous overload capacity of an electric motor.

- **service laterals**
  underground conductors supplying electrical service to individual residences and businesses.

- **service point**
  the point of connection between the electrical utility's wiring system and the individual customer's wiring system.

- **sickle**
  a sharp, curved metal blade fitted with a short handle. It is used to cut grain.

- **silage**
  feed resulting from the storage and fermentation of green crops in the absence of air.
- **silicon**
  a chemical element having four valence electrons. It is extensively used in manufacturing electronic components.
- **silicon-controlled rectifier (SCR)**
  a current-controlled solid-state switch that allows current flow in only one direction.
- **silo**
  an air-tight storage facility for silage.
- **simple circuit**
  an electrical circuit containing only one load.
- **single-phase electricity**
  a type of AC electrical service having one source voltage applied over two conductors.
- **single-pole circuit breaker**
  a type of over current protection device used to protect a circuit having one ungrounded conductor.
- **site specific crop management (SSCM)**
  the use of satellite, electronic, and variable rate technologies to automatically apply the correct production inputs to each specific field location.
- **sketch**
  a freehand drawing.
- **skirt**
  the bottom portion of a piston.
- **slag**
  the protective covering that forms over the weld bead when using certain welding processes.
- **sleeves**
  replaceable cylinder liners used on some internal combustion engines.
- **solar cell**
  a device for converting light into electricity through the photovoltaic effect.
- **soldering**
  a non-fusion joining of metals using a filler metal (solder) that melts at a temperature under 840°F.
- **solid state**
  a term designating electronic components without moving parts that can control current.
- **spark ignition engine**
  a type of internal combustion engine which requires an ignition spark to ignite

the air and fuel mixture at the beginning of the power stroke.

- **special purpose outlet (SPO)**
  a receptacle outlet that is installed for serving a specific electrical machine, appliance, or device.

- **specific gravity**
  the ratio of the weight or mass of a given volume of a liquid to the same weight or mass of the same volume of water.

- **speed**
  the rate at which a body moves.

- **split-phase motor**
  a type of squirrel cage AC induction motor developing fairly low starting torque while drawing a high starting current.

- **sprinklers**
  tools that spray water into the air to water crops. Its effect is much like rain.

- **squatting**
  settling on public or other unoccupied land.

- **squirrel cage rotor**
  a type of rotor that is commonly used in AC induction motors.

- **starting system**
  the engine auxiliary system that turns the engine crankshaft for engine starting.

- **starting winding**
  the auxiliary winding on the stator of an AC induction motor. It is used only when the motor is starting.

- **stator**
  stationary motor component consisting of a laminated steel core holding coils of copper wire in place.

- **stickers**
  crosspieces placed between layers of boards being air dried in order to allow air movement between layers.

- **strainer**
  a coarse filter that traps particles larger than the strainer openings.

- **stray voltage**
  small voltage difference existing between two surfaces, such as a metal waterer and the earth.

- **stress relieving**
  a form of annealing in which the metal is heated to just below its critical temperature, held for an extended period, and then cooled very slowly.

- **stroke**
  the movement of the engine piston from top dead centre (TDC) to bottom dead centre (BDC) or from BDC to TDC.
- **structural grade**
  lumber grade based on relative strength.
- **stud frame building**
  a structure made primarily of sawed lumber and constructed on a concrete or masonry foundation.
- **sub grade**
  a layer of earth and/or crushed stone and gravel levelled and graded for a foundation.
- **subsurface irrigation**
  water is supplied to the crop from underground pipes.
- **sulky**
  a light, two-wheeled carriage seating one person.
- **supercharger**
  a mechanically driven air pump used to increase the volumetric efficiency of an internal combustion engine.
- **surface irrigation**
  water flows over the soil surface to the crop.
- **sweating**
  1. the capillary action by which solder flows throughout a tight joint.
  2. the process of joining copper tubing by soldering.
- **switch**
  a device for opening and closing an electrical circuit.
- **switch loop**
  that part of a circuit from a light fixture located in the centre of the run to the switch located at the end of the run, controlling the light fixture.
- **synchronous alternator**
  device commonly used in electric power plants for generating electricity.
- **system grounding**
  the intentional connection of current-carrying portions of the electrical system to earth.
- **system objective**
  a statement describing the desired goal or outcome of a design project in measurable terms.
- **system operating parameter**
  the conditions under which the system objective must be met.
- **t**
  a sub-horizon symbol for illuvial clay. Accumulations of translocated silicate clay are indicated by the suffix t (Ger. ton clay). The suffix t is used

only with Bl as B2t to indicate the nature of the B.

- **T bar iron**
a rod or iron with a cross section in the shape of a T.

- **table**
1. a condensed collection and arrangement of related numbers of other items assembled in some definite order, as a table of logarithms
2. a surface, as a water table.

- **table system of sorting**
a system of sorting tomatoes, etc., in small canneries, in which the fruit is placed on a table and hand sorted for size, blemishes, etc.

- **table top terrace**
Bench terrace sloping outward or inward, with or without slight longitudinal grade, according as the rainfall of the tract is medium, poor or heavy, and the soil and subsoil are fairly absorptive or poorly permeable.

- **tableland**
a plateau or elevated region of flat or undulating country.

- **tabular drawing**
the drawing, either assembly or detail, on which the dimension lines are given reference letters and an accompanying table on the drawing lists the corresponding dimensions for a series of sizes of the structural parts or the machine parts, thus making one drawing serve the range covered.

- **tachometer**
1. in hydraulics, a device to show the swiftness of a current by its effect upon a submerged paddle or paddle wheel; a current meter having a rotating screw for driving clockwork indicating the speed in miles per hour.
2. an indicator, to show at a glance the speed of rotation.

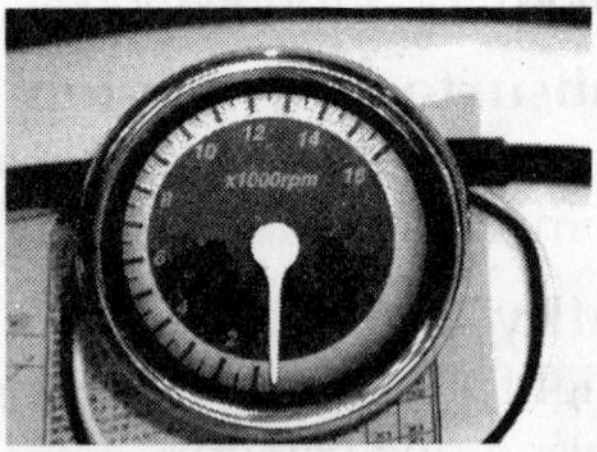

- **tack hammer**
a light hammer, used for driving tacks; instead of a peen, it is usually furnished with a clay for drawing them. The head is often magnetised, to hold the tack.

- **tack weld**
a preliminary weld used to align the pieces to be welded together. This promotes more uniform expansion due to heat of the final weld.

■ **tag line**
a small cable under tension which keeps the bucket from spinning swaying, twisting or drifting.

■ **tail**
1. the end of a canal.
2.the term is also applied to the work constructed at the end of a canal for the distribution of water.

■ **tail bay**
the part of a canal lock immediately below the tail gates.

■ **tail block**
a wooden pulley block which is provided with a tail or line for lashing it to any object, instead of the thimble generally provided.

■ **tail cluster**
two or more outlets at the tail of a distributing channel.

■ **tail control**
a device to control the water slope and level at the end of a model by either a mechanically operated gate or some improvised arrangement.

■ **tail cut**
cuts made at the lower end of a common rafter.

■ **tail drain**
a furrow or small ditch at the lower edge of a irrigated field for collecting surplus surface water for either redistribution or removal.

■ **tail escape**
an escape built at the tail of a canal to discharge surplus water to maintain the required full supply level, to scour the silt deposited in, and/or to empty, the tail reach.

■ **tail gates**
lock gates on the lower water level side of the lock.

■ **tail inlet transition**
the transitions provided upstream of a tail structure or at the tail end of, and contained within, the conduit length.

■ **tail outlet transition**
the transitions provided partly or wholly downstream of a tail structure or tail end of a conduit.

■ **tail pipe**
a suction pipe on a pump, used for irrigation, drainage or other purposes.

- **tail race**
  a channel conducting water away from a water wheel or water turbine.

- **tail reach**
  the last reach of a canal, i.e. the portion from the last control point to the tail.

- **tail regulator**
  regulator built at the outfall or tail of a waste channel.

- **tail tank**
  a reservoir supplied with water from a canal whenever in excess of canal requirements, having its own command and usually situated near the tail of a canal.

- **tail transitions**
  a transition or system of transitions provided at the tail end of a conduit or at a tail structure.

- **tail water**
  water level at the head of the watercourse or farm lateral just below the outlet.

- **tail water (hydraulics)**
  water in a river or channel immediately downstream of a structure; (irrigation) water that reaches the end of a crop field.

- **tail water rating curve**
  a curve showing relationship between the discharge and the corresponding tail water stage.

- **tail water recovery**
  the collection of tail water for reuse for irrigation.

- **tailboard**
  the gate at the rear end of a wagon or truck which can be let down or removed to facilitate loading or unloading. ,4/50 called end board, end gate, shoveling board.

- **tailing**
  the material which is unthreashed in the first time through the cylinder and concaves and which is recirculated into the thresher by means of the tailing elevator.

- **tailrace**
  a channel which conducts water away from a water-wheel; an after bay.

- **tainter gate**
  a crest gate whose face is a section of a cylinder, which rotates about a horizontal axis downstream from the gate, the water pressure against the gate being concentrated in the axis, reducing friction in raising and lowering the gate; a seal is placed

along sides and bottom of the gate face for water-tightness.

- **take off distributor**
device placed in the off take structure where volume of water to be diverted is only a small fraction of the water flowing in the parent channel, the volume drawn is generally constant in spite of the variations in flow and water levels in the parent channel. Except in small variations of water levels in the parent channel, the water level upstream of these take-off distributors is kept constant by automatic gates.

- **take up box**
in grain machinery, a device for taking up the slack on conveyer or elevator belts which is made in various patterns; the principle in each case being a shifting bearing for the staff of a conveyer pulley which can be moved to and fro in its frame by means of a screw to alter the tension of the belt.

- **taking off place**
a horizontal platform near the water surface level where the cattle or traffic get into the canal.

- **tal lands**
beyond the natural levees there are bowl shaped depressions geologically known as back water which though inundated are not subjected to erosion.

- **tall grass prairie**
the area in the western corn belt originally covered with tall native grasses, or other similar areas.

- **tall wall**
a wall of masonry or concrete built at the down-stream end of an outlet, containing the exit opening of the orifice discharging into the delivery channel. It protects the fill from scour and under-mining and serves as retaining wall.

- **taluds**
short, steep escarpments formed gradually at the down slope margins of fields by deposition against hedges or stone walls.

- **talus**
1. an accumulated heap of rock fragments derived from and lying in the base of a cliff or very steep slope.
2. a protection at the downstream end of a weir, fall, etc., consisting of blocks of concrete or masonry or stones.

- **talus spring**
a spring occurring at the foot of a talus slope, whose water

supply originated from water falling upon and seeping into the talus slope.

- **tampering**
the interference with the outlet structure or with its functioning by the water users.

- **tandem tractor**
two or more tractors with all front wheels removed attached together one behind the other and operated as a single unit or the front wheels of the front tractor may' be left in place and used to guide the tractors of a semi-tandem arrangement.

- **tangential flow**
in the case of a circular tank or a circular orifice, that flow which is in a direction perpendicular to the radii of the tank or orifice.

- **tangible losses**
flood damage that is susceptible to assessment in terms of money.

- **tank**
any artificial receptacle through which liquids pass or in which they are held in reserve or detained, for any purpose.

- **tank irrigation**
a system of irrigation in which small tanks or lakes are formed by putting up an earthen embankment along the lower run of a basin and which are fed by canal or natural catchment flow. The stored waters are led to the fields through sluices in the embankments and conveyed by a network of field channels.

- **tank sprayer**
any sprayer that is equipped with a supply tank or reservoir, as contrasted to the atomizer.

- **tank unit**
the unit of the fuel-indicating system that is mounted in the fuel tank.

- **tanker**
a specialised truck which carried a large, specially built tank for transporting of bulk milk.

- **tanking**
a method used to prepare livestock feed stuff from animal tissue.

- **tap**
1. in wiring, the connection of the end of one wire to some point along the run of another wire. There are many kinds of tap, as: Plain; aerial; knotted; cross (double and duplex; wrapped.)

2. a tool for cutting female or internal screw threads. It is made from a steel rod in which are cut the screw threads and has four flutings for clearance of chips in cutting.

- **tap borer**
  a boring instrument, shaped like the half of a hollow cone, used by coopers for making bung or tap holes in casks, or by plumbers for making or enlarging holes in leaden pipes.

- **tap root**
  a primary root which grows vertically downward giving off small lateral roots in acropetal suscession.

- **tap wrench**
  a key or spanner to operate a tap for threading holes. The usual pattern has a handle on either side of the socket for the square on the tap shank.

- **tape gage**
  a gage consisting of a tagged or indexed chain, tape, or other line attached to a weight which is lowered to touch the water surface whereupon the gage height is read on a graduated staff or opposite an index. Opposite ends of the tape may be attached to a float and counterpoise and the tape arranged to hang on a pulley, a similar reading arrangement being used.

- **tape sampler**
  a strip of white paper that catches dust from air drawn automatically through it at pre-set intervals. The result is a sequence of spots on the tape, made every hour or half hour. The relative blackness of the spots can be measured by a light meter (photometer) and they can also be chemically analysed.

- **taper drill**
  a long taper reamer or fluted cutter for forming conical holes under a drilling machine, long countersink.

- **taper pin**
  a conical dowel, fitting in a suitably shaped hole formed in two pieces of work, securing them together.

- **taper reamer**
  one designed to enlarge tapered holes; they are usually made to suit one of the standard tapers, either for taper pins, twist drills or the cock taper.

- **tapered aeration**
  the method of supplying varying amounts of air into the different parts of an aeration tank in the activated sludge process, more at the inlet, less near the outlet, and approximately proportional to the oxygen demand of the mixed liquor under aeration.

- **tapered roller bearing**
  bearings with tapered rollers designed to take both radial and end thrust loads; the relative capacities depending upon the amount of taper.

- **tapered single side main duct system**
  a system of ducts used in barn drying hay. In this system there is one main duct along one wall, which gradually diminishes in size from the end the fan is on. The lateral ducts may or may not diminish in size.

- **taphoon**
  a portable dam or stop in the form of a canvas on a frame or metal panel put across ditches when it is desirable to use ditches in sections, filling new parts lower down as the irrigation progresses.

- **tapoon**
  a method of measurement of seepage losses from canals in which a selected canal section is segregated by dikes at each end. Seepage losses are most accurately determined by measuring the inflow required to maintain the water level in ponded section at a given level, due allowance being made for evaporation.

- **tappet clearance**
  the clearance between the end of the valve stem and the valve lifter.

- **tartk truck**
  a truck or trailer which is equipped with a large tank to transport liquid products, such as milk.

- **tempering**
  a heat treating process that reduces the hardness of a hardened metal by relieving a desired amount of internal stress.

- **template**
  an outline pattern from which letters and symbols can be traced.

- **tensile strength**
  the ability of a material to withstand a pulling force.
- **terminal board**
  the board at which electrical connections are made to an electric motor. It is usually located in one of the end bells.
- **thermal conductivity**
  the relative ability of a material to transfer heat.
- **thermal efficiency**
  the ratio of the mechanical power produced by an engine to the amount of heat energy consumed.
- **thermal protection**
  a motor overload protection technique relying on heat production to open the motor circuit in case of an overload.
- **thermocouple**
  a device consisting of a junction of two dissimilar metals that facilitates the generation of voltage when the junction is heated.
- **thermodynamics**
  the branch of physics dealing with heat flow.
- **thermopile**
  an electrical-generating device consisting of a number of thermocouples connected in series to increase output voltage.
- **thread pitch**
  refers to the spacing and size of the threads on threaded connectors, taps, and dies.
- **three-phase electricity**
  electrical service consisting of three interrelated voltages for which the phase difference is 120 degrees.
- **threshing**
  removal of grain from the plant.
- **throw**
  refers to the number of paths provided for current to flow out of a switch.
- **thyristor**
  a class of semiconductor switching device.
- **tillable**
  soil that is workable with tools and equipment.
- **time-delay**
  a fuse or circuit breaker that will tolerate a small, temporary overload without opening the circuit.
- **tinning**
  the process of applying a very thin layer of solder to a base metal.
- **top dead centre (TDC)**
  piston position when the piston is at the top of its stroke

or when combustion chamber volume is at a minimum.

- **torque**
a twisting or rotating force.

- **toughness**
measure of a metal's resistance to impact.

- **tractor**
source of power for belt-driven machines as well as for pulling.

- **transistor**
a semi-conductor device used for amplification and switching applications in electronic circuits.

- **triangle**
mechanical drafting tools used either alone or in pairs for producing a variety of angles.

- **triplex cable**
a cable consisting of three conductors twisted together. It is commonly used for overhead service drops.

- **trivalent element**
a chemical element having three valence electrons.

- **true power**
the amount of electrical power actually consumed.

- **T-square**
a mechanical drafting tool that forms a 90-degree angle with the left or right edge of the drawing board.

- **tungsten inert gas welding (TIG)**
electric arc welding process that uses a non-combustible tungsten electrode, an inert shielding gas, and possibly a non-coated filler rod. It is also called gas tungsten arc welding (GTAW).

- **turbocharger**
a turbine-type air pump, driven by engine exhaust gases, designed to increase the volumetric efficiency of an internal combustion engine.

- **Type S fuse**
a type of fuse designed to prevent 'over-fusing' a circuit. It is also called a 'non-tamperable' fuse.

- **u**
1. symbol for the specific surface area of the soil.
2. symbol pertaining to subtypes Ui weakly evolved; U_2 moderately evolved; U_3 completely evolved; U, degraded.
3. symbol for uniform brown soil.

- **u abutment**
form of abutment where the wing walls return at 90°.

- **u dometer**
a rain gauge.

- **ukra microscope**
microscope used for seeing small particles in liquid by means of a strong light coming in from the side and causing the particles to be seen as bright points.

- **ultimate analysis**
the determination of the elements in a substance or mixture, as a soil, in contrast to proximate analysis where the compounds in a mixture are determined.

- **ultimate bearing capacity**
the average load per unit of area required to produce failure by rupture of a supporting soil mass.

- **ultimate compressive stress**
used in reference to concrete. It means the most strength the concrete will gain which occurs 28 days after the start of curing.

- **ultimate particles**
soil particles after a standard dispersing treatment, with all aggregates broken down.

- **ultimate pH**
the pH after all diffusible anions and cations have been removed.

- **ultimate set**
the difference between the length of a specimen plate, before testing and at the moment of fracture, expressed in percentage of the length.

- **ultimate strength**
unit stress when a material is carrying its maximum load before rupture.

- **ultimate wilting point**
the moisture content of the soil at which all the leaves of plants growing in that soil are completely wilted and will not recover in an approximately saturated atmosphere without addition of water to the soil. The lower end of the wilting range.

- **ultisols**
soils of humid areas characterised by the presence of either an argilic horizon or a fragipan each of which is less than 35 per cent saturated with bases. These soils are usually moist, but during warm season some turn dry.

- **ultra-filtration, hyper-filtration**
  the use of filters capable of selectively sieving out particles smaller than 10 tim down to 0.002 run. To prevent the membrane blocking too rapidly with built-up solids, the water flows past the upstream side of the membrane, sweeping most of the solids with it in the more concentrated waste stream. The membranes are made up of polysulphone or polyacrilonitrile or cellulose acetate, and need frequent, laborious cleaning. The operating pressure can be between 1 and 7 atm.

- **umbric horizon**
  a thick dark-coloured surface horizon (epipedon) similar to a mollic horizon but with low base saturation and may be highly H saturated.

- **unaccounted for water**
  that portion of the water supply taken from a source into a distribution system, which is not delivered to the consumers, or otherwise accounted for by municipal usage.

- **unavailable soil moisture**
  soil moisture held so firmly by molecular forces that it cannot ordinarily be absorbed by plant roots with sufficient rapidity to produce growth.

- **unavailable water**
  water held by the soil so strongly that root hairs cannot readily absorb it.

- **unavoidable farm losses**
  principally those losses which are due to normal field evaporation and deep percolation.

- **unbalanced vane pump**
  a type of hydraulic vane pump in which the pump rotor and drive shaft are subjected to unbalanced forces.

- **uncased well**
  a tube well, such as in deep rock, with little or no casing of any kind.

- **uncertainty principle**
  the indeterminacy principle of Hedisisenberg which postulates that any physical measurement turbs the system being measured, thus limiting the accuracy of measurements, especially at the atomic, nuclear and particulate levels of magnitude.

- **unconfined aquifer**
  a permeable bed only partly filled with water and overlying a relatively impervious layer. Its boundary is formed

by a free, water table or phreatic level.

- **unconfined ground water**
ground water not under hydrostatic pressure, its water table rising and falling, depending upon the change in the volume of water stored.

- **unconformity spring**
a contact spring occurring at the surface outcrop of the boundary between two different geologic formations, the upper one being water bearing and the lower one being impervious or less pervious.

- **unconsolidated undrained test**
a soil test in which time is not allowed for consolidation and the water content of the test specimen remains practically unchanged during the application of the confining pressure and the additional axial (or shearing) force.

- **uncontrolled mosaic**
a mosaic in which the photographs have been adjusted by reference to ground control.

- **uncontrolled spillway**
a spillway, the crest of which permits water to discharge, whenever the reservoir surface is higher than the crest.

- **uncontrolled storage**
reservoir storage not controlled by gates or other devices.

- **undamped natural frequency**
the frequency of free vibration resulting from only elastic and inelastic forces of the system.

- **under consolidated soil**
a deposit that is not fully consolidated under the existing overburden pressure.

- **under inlet**
any device other than a well which admits ground water to an underdrain. In its usual form it consists of an opening filled with broken stone or gravel.

- **undercoaring**
the process of spraying of a plastic material on the exposed underparts of an automobile to slow down corrosion and to deaden noise.

- **undercut slope bank**
the bank of a meandering stream which is eroded by stream action.

- **undercutting**
cutting away at the base, or underpart, of a steep slope or

cliff, as by stream glacier, wind, or wave erosion, thereby steepening the slope or producing an overhanging cliff.

- **underdrain**
arrangement for draining away excess water, with the drain or trench dug deep into the soil and covered up so that the soil could be cultivated without interruption and the profile drained.

- **underdrainage**
that drainage consisting of drain tiles placed in trenches deep enough so that the covering soil can be cultivated and the profile adequately drained.

- **underdrained**
1. ground provided with an underdrain or a system of underdrain is said to be underdrained.
2.land is also said to be underdrained or under designed if tile laterals are spaced wider than desirable.

- **underdrained sand filter**
a method of disposing of sewage by filtering it through sand after it leaves a septic tank.

- **underdrains**
pipes under a trickling filter or true filter, which remove the water, in the rapid gravity sand filter, dual media filter and multi-media filter they also provide access for the backwash water.

- **underfire air**
the air passing through the grate of a furnace. It should be kept to a minimum so as to reduce the carryover to the dust arrestors.

- **underflow**
movement of water through a pervious subsurface stratum (underflow conduit), possibly under the bed of a river. In general it follows the same course as the surface drainage.

- **underflow conduit**
a permeable formation which underlies a surface stream channel, whose bottom and sides are more or less definitely limited by formations of relatively low permeability, and which contains ground water which percolates downstream.

- **underflow density currents**
the density currents flowing under another fluid.

- **underflow turbidity currents —settling type**
in this case, the inflowing muddy water does not dive

under the clear water, but pushes it downstream and forms a considerable body of muddy water, in which the sediment slowly settles to the bottom, which later on flows down into the reservoir as in the form of underflow turbidity currents. The existence of these currents is indirectly inferred from the deep deposits of fine material in the streambed near the upper ends of some large reservoirs.

- **underflow turbidity currents—plunging type**
in this case, the turbid water flowing into a reservoir plunges directly under the clear water in the lake. The formation of this type is indicated by the collection of flowing drift on the surface and the sharp line of separation of the muddy and the clear water.

- **underflowing turbidity currents**
it is the common type of density currents, which move down the bottom of a reservoir, dyke or deep channel due to the greater density of the turbid water resulting from the inclusion of suspended sediment.

- **underground reservoirs**
saturated coarse gravel wells beneath the ground surface.

- **underground runoff (seepage)**
water flowing toward stream channels after infiltration into the ground.

- **underground utilities**
systems of pipes and conduits beneath the surface of city streets or developments for conveying water, sewage, drainage, gas, etc., and for the protection of electric wires, etc.

- **underground water**
these terms are sometimes used as synonymous with ground water and sometimes as synonymous with subsurface water.

- **underground water basin**
a valley area or plain underlain by ground water which has established an operational regime from the prevailing hydrologic and climatic conditions to behave as a unit and which can usefully be developed as a unit.

- **underground watercourse**
a geological formation which contains water flowing in a

known and defined channel. Rights to water in underground watercourses are in most states similar to rights to water in natural surface watercourses.

- **underirrigation**

1. supplied with water for crop production entirely or in part by means of irrigation.

2. supplied with an inadequate amount of irrigation water.

- **underreamer**

in well boring, a tool having two legs or cutters, spread open by a spring, which is used to ream or enlarge the borehole underneath the casing.

- **underrunner**

a grinding mill in which the lower of the two stones revolves, the upper remaining stationary. This form is especially useful in grinding oats, corn, chalk, clay, etc. The material rests on the revolving stone, and is moved out from the centre by centrifugal force.

- **undershot head gate**

a canal head gate, where the water passes from the river into the canal through gate openings, formed when raising the gates between the sill of the gate opening and the lower edge of the gate.

- **undershot type regulator**

a regulator where the water passes from the parent channel to the off taking channel below the gate opening formed (when raising the gates) between the sill of the gate opening and the lower edge of the gate.

- **undershot wheel**

a water wheel operated by the impact of flowing water against paddles or vanes attached to the periphery, these being partly or wholly submerged in the moving stream of water.

- **undersluices or scouring sluices**

1. a portion of a weir or barrage, having its sill at a level lower than the remaining portion; provided (1) adjacent to the head regulator and or (2) at the approximate location of the line of maximum surface velocity before the erection of the weir or barrage with a view to sluicing out detritus

during periods of high flows in the river.
2. the term 'undersluices' is also used for 'sliding sluices' in main dams, where water is discharged at a level much below the reservoir level.

- **undersluices pocket**
the portion of river channel upstream of the undersluices bounded by the divide wall, inlet of the head regulator structure and the undersluices.

- **Underwriters' Laboratories, Inc (UL)**
a nationally recognised non-profit organisation providing a voluntary product testing and listing programme for electrical (and other) materials and devices.

- **undifferentiated soil group (mapping unit)**
two or more soil or land types which may be mapped as one unit, as their differences do not influence soil management or are not relevant to the purpose of soil survey.

- **undisturbed sample**
samples in which the material has been subjected to so little disturbance that it is suitable for all laboratory tests and thereby for approximate determination of the strength, consolidation, and permeability characteristics and other physical properties of the material *in situs*.

- **undulating**
the lay of the land with elevations and depressions; not flat or level or even; wavy margins of leaves.

- **undulating topography**
land that is gently sloping in different directions. Slopes are not steep enough to interfere with easy cultivation.

- **undulation**
a continuously propagated motion to and fro, in any fluid or elastic medium, with no permanent translation of the particles themselves.

- **undulatory movement**
a rising and falling as if on waves.

- **unfilled porosity**
the fraction of the soil space occupied by the gaseous phase alone.

- **unflumed fall**
a fall or drop in which the throat width is not less than the designed bed width of the channel downstream of the fall.

unfree water
water which is held closely by the soil colloids so that it is not readily available to plants. It freezes well below 0°C.

ungrounded conductor
a circuit conductor that is not attached to the ground. It is commonly called a 'hot' conductor.

uniform channel
a channel whose cross-section does not change along its length, is straight in alignment and has a constant grade.

uniform flow
the flow of a liquid in an open channel is said to be uniform when the depth and the other elements of flow, such as the cross-sectional area, the velocity and the hydraulic slope, remain constant from section to section. The surface line in this case is parallel to the bottom line.

uniform pitch
in mechanics, a screw is said to be of uniform pitch when there is an equal distance between the helices, or blades, as distinguished from increasing pitch.

uniform soil
soil in which only one particle size extends throughout one bed or layer of soil.

unit factors
ratios whose actual value equals unity, or one.

United States Customary System (US)
the system of measurement commonly used in the United States based on the pound, foot, and gallon as the basic units of weight, length, and capacity, respectively.

universal motor
a type of wound-rotor electric motor that can operate on AC or DC electricity.

unloading valve
a type of hydraulic valve which by-passes pump output to the reservoir when a set pilot pressure is maintained.

upper plumb cut
a cut made at the top end of a rafter. It is also called the ridge cut.

use of water
a legal term to designate the beneficial purpose for which water is diverted from a stream or other body of water.

- **useful storage**
storage between maximum and minimum operating levels.

- **useless water**
water which cannot be absorbed by plant roots, or which harms them, e.g. hygroscopic water and gravitational water.

- **utilisation (forestry)**
the branch of forestry which deals with the harvesting, marketing, conversion and putting the forest produce for a variety of uses, e.g., timber, fuel, charcoal, pulp wood, plywood and a host of other products.

- **utilisation coefficient of soil's reserve**
the ratio of the quantity of a nutrient substance present in the soil to that withdrawn by the plant.

- **utilisation factor**
1. the ratio, usually expressed decimally, of the water flowing in a stream which is used for power development, to the quantity which is available for use; the latter is limited by the flow as indicated by the flow-duration curve, and also, at higher stages, by the capacity of the water wheels.
2. the factor representing the ratio of net to gross power.

- **utility building**
a farm building which is designed to house miscellaneous equipment, machinery and supplies.

- **V belt**
belt made of rubber and other reinforcing materials that fit into pulleys which have V-shaped grooves.

- **V block**
in machine shop practice, a carefully planed block of steel or cast iron with a V on one side, intended to be used in pairs to support shafts or other cylindrical objects on the marking of plate, or, on the table of a planning machine, etc.

- **V ditcher**
a wooden tool used for construction or cleaning of field channels for irrigation or drainage; also used for making border ridges or bunds.

- **V drag, or V crowder**
a V-shaped wooden or steel implement for construction and maintenance of small surface ditches. In operation, the operator stands on the runner board and the position taken

by him controls the depth of digging.

- **V engine**
a type of engine in which the cylinders or rows of cylinders are placed at an angle forming the letter V.

- **V flat drive**
the driving of a relatively large flat pulley with V-belts from a smaller, grooved sheave.

- **V gear wheel**
in machinery, a duplex arrangement of skew gearing in which each tooth is of a shape of the letter V.

- **V hook**
an eccentric rod grab having V-shaped jaws.

- **V notch weir**
a triangular notch in a wall built across a stream used to measure the discharge of the stream.

- **V shaped ring**
a V-ring; a gasket ring with cross-section in the shape of a 'V' which is used on pistons and plungers for sealing against liquid and pressure losses in pumps and hydraulic equipment.

- **V thread**
a V-shaped, screw thread.

- **vacuum breaker**
a device for relieving a vacuum or partial vacuum formed in water supply lines to prevent backsiphonage.

- **vacuum distillation**
process of distillation at low pressure and temperature used for distilling ammonia in the presence of urea.

- **vacuum dryer**
a device used for evaporating moisture from a product in partial vacuum. In milk processing, this device is commonly used for the dehydration of milk where a coating of milk is spread on a revolving steam heated drum located in a vacuum and is scraped off as a solid.

- **vacuum filter**
a horizontal shaft, continuously rotating drum with filter cloth or metal coil stretched around it, subjected vacuum inside the drum.

- **vacuum gage**
a diagnostic tool used for testing the engine intake system.

- **vacuum gauge**
instrument to measure the unbalanced pressure of the atmosphere upon condensers etc.

vacuum oven method
a method of moisture determination for grain.

vacuum pump
1. a pump for creating a partial vacuum in a closed space.
2. a pump in which water is forced up a pipe by the difference of pressure between the atmosphere and a partial vacuum.

vacuum tank
closed unit in which the pressures can be maintained to any desired degree of vacuum so that cavitations effects in a model can be observed readily.

vadose water
1. subsurface water intermediate between suspended soil water and fringe water. All subsurface water in the soil from the water table to the ground surface.
2. water in excess of adsorption water and pellicular water seeping towards the water table.

vadose zone
the zone of aeration.

valence ring
the outermost electron orbit of an atom.

valley
1. an elongated depression, usually with an outlet, between bluffs or between ranges of hills or mountains.

valley fill
the sedimentary deposits laid down in a valley subsequent to the time of its formation.

valley spring
a spring occurring on the side of a valley at the outcrop of the water table.

valley storage
1. the volume below the water surface profile.
2. the natural storage capacity or volume occupied by a stream in flood after it has overflowed its banks. It includes the quantity of water within the main channel (channel storage) and that which has overflowed its banks (lateral storage).

valve
a device installed in a pipeline for the purpose of controlling the magnitude and direction of the flow. It consists essentially of a shell and a disk or plug fitted to the shell.

valve chamber
a structure, housing the regulating valves, control mecha-

nisms for operation of valves, equipment required to remove parts for repair, etc.

- **valve clearance**
the gap between the valve tappet and the valve stem.

- **valve controlled hydrant**
in irrigation, a hydrant fitted with a small valve which is connected by a short riser to the underground lateral.

- **valve elbow**
a fitting installed in a pipeline system to allow sprinkler lines or laterals to take off in any direction; rotates 360°.

- **valve face**
the angled surface of the engine valve that seals against the valve seat.

- **valve gear**
the mechanism to control the motion of the valves admitting and exhausting to or from the cylinders of an engine.

- **valve grinding compound**
very fine emery powder mixed with grease or any other viscous lubricant meant for grinding valve seals is called valve grinding compound.

- **valve lifter**
the cylindrical part of the engine that rests on a cam of the camshaft and is lifted by the cam action so that the valve is opened. There is a valve lifter for each valve.

- **valve rope knife**
an emergency tool which is used in well boring on the cable system consisting of a cylinder with a knife balanced in it like a valve in a pump.

- **valve spring**
a spring which fits around each valve to force it to return to a closed position each time after being forced open by the valve lifter.

- **valve stem**
the rod by means of which a valve is opened or closed, the rod lifting and pushing down the gate in the operation.

- **valve tower**
a hollow cast-iron or masonry tower built within a reservoir and equipped with draw off pipes at different levels for taking off water for supply purposes.

- **van't hoff's law**
the osmotic pressure of a dilute solution is equal to the pressure which the solute

would exert in the gaseous state, if it occupies a volume equal to the volume of the solution at the same temperatures.

- **vane**
  a weathercock for indicating the direction of wind.
- **vane axial fan**
  a type of axial flow fan in which air moves through the fan parallel to the impeller shaft. It is distinguished from other axial flow fans by large buds, and short blades plus stationary straight vanes on the discharge side of the fan to prevent rotation or swirling of the air leaving the fan.
- **vane type pump**
  a pump that consists of a circular impeller set off centre in a body casting, with two vanes or wings.
- **vapour**
  moisture in the air; gaseous form of a normally liquid or solid substance; kinds of remedial agents to be inhaled.
- **vapour barrier**
  a sheet-type, or air film material which is impervious to the passage of water vapour under normal lines are concentric circles and in which the total head for each stream line is the same.
- **variable capacitor**
  a type of capacitor whose capacitance value can be adjusted.
- **variable displacement pump**
  a type of pump in which the volume of fluid pumped per cycle can be changed while the pump is in operation.
- **variable rate technology (VRT)**
  automatically varying the rate of application of production inputs, such as feed, seed and fertiliser, without stopping.
- **variable resistor**
  a type of resistor in which the resistance value can be adjusted.
- **velocity**
  speed or quickness of motion.
- **veneer**
  a thin layer of wood that is obtained by peeling a continuous layer of wood from a log.
- **viscosity**
  a measure of a liquid's resistance to flow.
- **volt (V)**
  the unit of electrical pressure, or electromotive force.

- **volt meter**
an instrument used for measuring voltage.

- **voltage drop**
the reduction in voltage that occurs as electricity flows through a resistance.

- **voltage source**
the origin of the electromotive force in a circuit. For example, a battery or alternator.

- **volumetric efficiency**
a measure of the air-pumping ability of an internal combustion engine.

- **vughs**
these are unconnected voids with irregular shape and irregular walls. In general they are found in soil material that has a high proportion of fine textured material capable of strong cohesion and adhesion. The vughs may be formed as the fine material cements or aggregates the primary particles.

- **wadder**
the wrapping of rope round the collar of the rope spear, or screwing an iron disk on the spear collar in order to cause the rope to pack in the bore hole to give the spear something to grab.

- **wage**
the amount paid periodically to the labourer for the work done in the field.

- **wagon box dryer**
a grain or hay heated air dryer for drying the product while on a wagon. A canvas is placed over the wagon and heated air is either forced up under a perforated floor or down through the product and the floor.

- **wagon drag**
a skid, shoe, or drag which is used to check the speed of a wagon on a downgrade.

- **wagon drill**
a drifter mounted on a mast supported by two or more wheels; hence 'wagon drilling'.

- **wagon hay loader**
a machine attached to the rear of the wagon and driven over the windrow or along the swath from which it gathers hay, loading it on to the wagon.

- **wagon hitch hinge**
the type of hitch hinge which is used on a one-row pull type corn picker. This hinge supports the hitch at the same elevation as the tractor drawbar to provide for easy connections.

- **wagon hots**
a lifting device for wagons for unloading into elevators or conveyors.
- **wagon jack**
a lift used to raise the axle of a wagon, etc., sufficiently to admit removing a wheel for axle greasing or repairs. A device to lift one end of a wagon to facilitate unloading.
- **wagon spout**
the spout at the end of a corn picker elevator which directs the snapped corn downward into the wagon. Also used on a forage chopper.
- **wakefield piling**
wide boards or planks so nailed together that the centre is offset to form a tongue and groove. When driven the tongue of one pile rides in the groove of the adjacent pile.
- **walking beam**
a heavy hinged arm with a sheave at one end through which the drilling line passes; it provides for the raising and dropping of the tools in percussion drilling where drilling rigs are equipped with a walking beam and not a spudding beam.
- **walking plough**
the first type of plough developed successfully. It is horse-drawn and is called walking plough because the operator walks behind the plough.
- **walking type tractors**
power tillers used for ploughing, harrowing, levelling, puddling, weeding, harvesting, transporting; also for stationary jobs; 3 main types, rotary, crank and screw types; have 2 to 3 H.P., aircooled, single cylinder, fourcycle gasoline, or kerosene engines; number of attachments used.

- **wall bearing**
in machinery, a bearing for receiving a shaft when entering or passing through a wall.
- **wall drill**
a portable drilling machine without pillar or standard which may be attached to a wall or to any upright post for temporary work, or to suit special circumstances.

- **wall friction**
frictional resistance mobilised between a wall and the soil in contact with the wall.

- **warm house**
a house or unit of a range of greenhouses which is maintained at a day temperature of about 75° and night temperature of about 60°.

- **warm spring**
a thermal spring the water of which has a temperature lower than that of the human body, that is, 98 degrees F.

- **warping**
the process of causing a deposit of silt on land which generally becomes fertilised by such deposit.

- **warren hoe**
a garden hoe with a heart-shaped blade which is especially used for making furrows.

- **wash drill**
a device for boring holes through soft material. The drill point is raised and lowered, churning and loosening the material at the bottom of the drive pipe. Water is forced through the hole in the drill point and flows upward and out of the top of the drive pipe, washing the cuttings out.

- **wash ins**
holes formed in the land where irrigation water flows downward through the backfilled trench, washing large volumes of soil into the closed drain through the joints.

- **wash leather**
yellow, soft, oiled sheepskin leather used for polishing metal.

- **wash load**
suspended material of very small size (generally clays and colloids) originating primarily from erosion on the land slopes of the drainage area and present to a negligible degree in the bed itself.

- **wash water**
water which is used to wash the filter beds in a rapid sand filter.

- **wash water recovery tank**
a tank where dirt in the wash water settles out and the top water is recovered to pass through the water treatment plant again, while the sludge is removed for dewatering.

wash water trough
troughs are provided above the filter and/or anthracite in modern rapid or multi-media filters, etc., to remove wash water. The troughs should be about 0.5 m above the level of the medium when it has settled. This allows for expansion of the bed during backwashing and ensures that sand is not lost with wash water.

washing nozzle
nozzle at the end of the jetting pipe through which water shoots and then rises in the annular space between the casing and the jet tube, carrying with it any contaminated material and finally over-flowing into the settling tank at ground level.

washout
the washing out or away of earth from the canal banks, below hydraulic structures, caused especially by accumulation of debris, burrowing animals, erosion and bad drainage of the subgrade, leading to complete breakdown of the banks, structures, or lining.

washout valve
a valve provided at a low point in a main which allows debris to be flushed out at a point where the waste water can conveniently drain away.

waste
desolate, desert, uncultivated, as a result of natural barrenness or other natural causes; unless remains, refuse, scraps, shreads; any part of a produce that is not put to use.

waste bank
the ridge of excavated earth parallel to a drainage channel.

waste channels
channels which are extensions of the distribution system and convey surplus or unused water to a natural after course, depression or other point where the waste or unused water can be disposing of.

waste gate
a device used for controlling the amount of boost produced by a turbo-charger.

waste stabilisation pond
the five types of lagoon for sewage treatment are discussed under aerated lagoon, anaerobic lagoon, facultative lagoon, high rate aerobic lagoon and maturation pond.

waste water
the water (excluding that spent in conveyance losses)

which unavoidably runs surplus from out of water for beneficial uses due to breaches, closures or other mishaps.

- **waste weir**
  a structure in a canal or open conduit installed to allow spilling or wasting of excess water from other sources which might find its way into the canal and cause damage by overflowing the banks. It usually consists of a short section of canal with a level crest, protected against erosion or cutting, the elevation of which is below the top of the canal bank but above the normal water surface in the canal.

- **wasteways**
  structures provided in canal systems to dispose of all excess water, or empty a canal.

- **water account**
  an account maintained of distribution of supplies between units of interlinked canals or different components of one canal system.

- **water analysis**
  the determination of the physical, chemical and biological characteristics of water. Such analyses involve usually four different kinds of examinations: bacterial, chemical, microscopic, and physical.

- **water application efficiency**
  the ratio of the volume of water that is stored by the irrigator in the soil root zone and ultimately consumed (transpired or evaporated, or both) to the volume of water delivered at the farm.

- **water application rate**
  the rate in inches per hour that irrigation water is applied to fields.

- **water balance**
  it is hydrological balance between rainfall plus irrigation to that of changes in soil moisture, evaporation, percolation and runoff.

- **water balance equation**
  P=Q+U+E+W, where P is the precipitation or irrigation water added, Q is the run-off, U is the deep drainage passing beyond the root zone, E is the actual evapotranspiration, and W is the change in soil water storage.

- **wattmeter**
  an instrument used for measuring the power in an electrical circuit.

- **watts (W)**
the basic unit of electrical power.
- **way**
refers to the number of paths for oil flow through a directional control valve, including reverse flow.
- **welding**
the process of joining materials by fusion.
- **whole systems planning**
a planning system in which specific decisions are made based on a holistic, global perspective.
- **wire nut**
a common type of connector for making splices without solders in electrical conductors.
- **wire stripper**
an electrical wiring hand tool for removing insulation from individual conductors.
- **wire wrapping**
a technique for assembling permanent electronic circuits that does not involve solders.
- **with diversity**
electrical loads in a building or other load centre that are not expected to operate at the same time as the loads without diversity in the same building.
- **without diversity**
electrical loads in a building or load centre that, under normal conditions, all operate at the same time.
- **work**
the movement of a force through a distance.
- **wrench**
tool used for tightening nuts. Pipe-wrench-tool for taming a pipe round so as to screw it into another pipe.
- **wrist pin**
the pin which connects the pitman of a cutter bar to the pitman wheel. It has a flattened head to secure the pitman. In machinery, a stud or pin which forms a journal in a crosshead or piston.
- **wrought iron pulleys**
in machinery, these have largely superseded cast iron pulleys, because of their superior lightness and reliability. The rims are formed of wrought iron or steel and the arms are of the same material, each being made distinct from the other; they are put together in halves and are, therefore, split pulleys.

■ x
a symbol for fragipan character used as a suffix with horizon designation to indicate genetically developed properties of firmness, brittleness, high density, and characteristic distribution of clay that are diagnostic of fragipans. Fragipans or parts of fragipans, may qualify as A2, B, or C. Such horizons are classified as A2, B, or C, and the symbol x is used as a suffix to indicate fragipan character. Unlike comparable use of supplementary symbols, the symbol x is applied to B without the connotative Arabic numeral normally applied to B. Arabic numerals used with C to indicate only vertical sub-division of the horizon precede the x in symbol, as Clx, C2x.

■ xeric
dry; pertaining to soils which normally have a low moisture content, and to plants which tolerate a very dry soil or arid conditions.

■ xeric environment
an environment in which the soil contains very little water, and the atmospheric conditions favour the rapid loss of water from plants. A soil moisture regime typical of mediterranean climates having moist cool winters and warm dry summer. Irrigation or summer fallow is generally essential for crop production.

■ xeromorphe
a soil the principal conditions of whose formation are through to have been great dryness of the climate or of the soil, with predominance of evaporation.

■ xeromorphy
having morphological characters which aid in the reduction of water loss.

■ xerophile
a plant which tolerates a dry habitat.

■ xerophytes
plants that grow habitually in dry sites (soils), where the evaporation stress is high and the water supply is low.

■ xerophytic vegetation
vegetation, characteristics of the desert regions, such as thorny brush, cacti, shrubs, and small flowering annual and perennial plants.

■ xerosere
a plant succession or sere having its origin in a dry soil.

- **xerothermic index**
an index of the relationship of drought to plant growth.

- **xerothermique**
the number of days without rainfall, corrected taking into account the days of fog, of dew, and relative humidity according to a certain scale.

- **X-ray crystallography**
the use of X-rays to study crystal structure. The way in which X-rays are diffracted gives information about the spacing between crystal planes.

- **X-ray diffraction**
a means of studying clay mineralogical composition.

- **X-ray diffraction analysis**
a means for determining the internal structure of a substance by the diffraction pattern formed when a pin-point beam of X-rays passes through it.

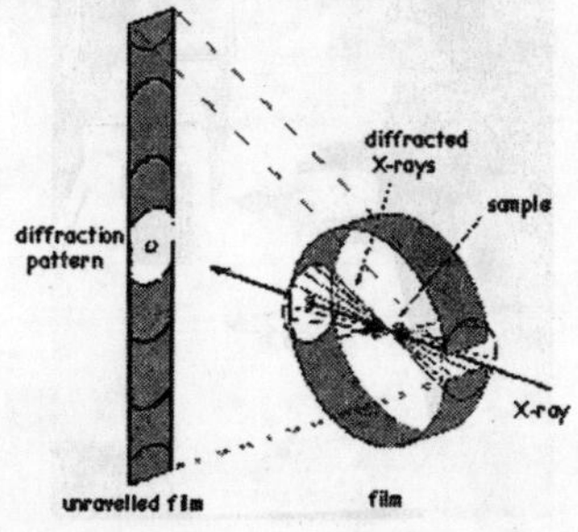

- **X-ray emission spectrograph**
an analytical tool that provides a non-destructive means of determining elemental concentrations in plant materials with minimum sample preparation.

- **X-spring**
a carriage spring, composed of two superimposed laminated springs, forming a letter X.

- **X-unit**
a unit used to express the wavelengths of X-rays.

- **xylem**
the softwood that is located between the inner bark of a tree and the heart-wood. It is also called sapwood.

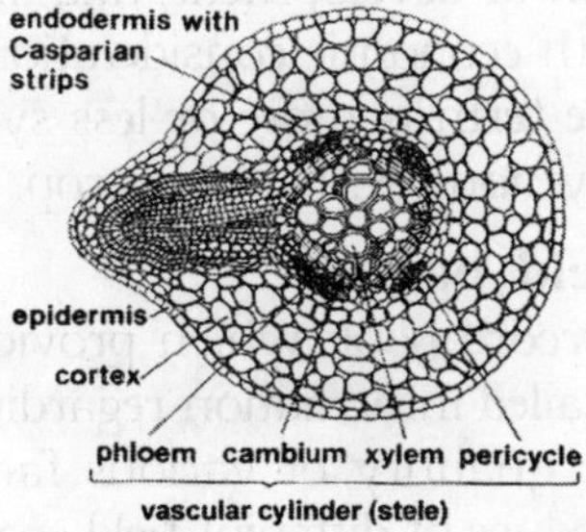

- **y-divider**
a structure in the form of Y to divide flow into two ditches. The channel of approach must provide parallel flow without cross-currents or other ob-

structions, and the flow of the channel just above the divider must be set level transversely.

- **yellow earths**
a group of soils in the tropics, and warm temperature regions. The yellow colour is probably connected with a high degree of hydration of the ferric oxide.

- **yield**
the quantity of water expressed either as a continuous rate of flow (cubic feet per second, etc.) or as a volume per unit of time (acre-feet per year, etc.) which can be collected for a given use, or uses, from surface or ground-water sources on a watershed. The yield may vary with the use proposed, with the plan of development, and also with economic considerations. The term is more or less synonymous with water crop.

- **yield book**
a record book which provides detailed information regarding the quantity of various farm produce of different field crops grown on the farm with their income through sale on acre basis.

- **yield capacity**
the total quantity of material per annum, of given species, that an area is capable of producing under normal conditions.

- **yield components (yield attributes)**
are components which finally make up or control yield of any crop for example yield components cereal grain yield are grain weight per grain, grain number/ear, ear number/plant and plant number/unit area.

- **yield determinant**
any factor, natural or due to human intervention, influencing crop yield.

- **yield maximisation**
agronomic practices adap-ted to get highest possible crop production per unit area per unit time without considering either the cost of production or net return.

- **yield of drainage basin**
total volume or flow from a drainage basin for a long stipulated period of time, e.g., annual yield of drainage basin is the mean annual runoff.

- **yield point**
the point at which a metal on being stretched becomes permanently deformed.

- **yield stability**
the maintenance of yield at desired level over a period of time. Its index is measured by inverse of coefficient of variation, regression between yield and environmental index and disaster level estimated by probability of a system.

- **yield sustained**
continual annual or periodic yield from plants in an area through management practices which help to maintain productive capacity of the land.

- **yield table**
a tabular statement which summarises on an unit area basis, all the essential data relating to the development of a fully stocked and regularly thinned at periodic intervals.

- **yield, n-years dependable**
the minimum supply of a given water development that is available on demand, with the understanding that lower yields will occur once in n-years, on the average.

- **yoke**
1. a variously shaped wooden frame which is placed on the necks of a team of oxen to which the traces are attached.
2. a clamp which unites two pieces.
3. a wood or steel bar placed at the end of a tongue connected by rings and leather straps to the names of the harness of a pair of horses. It holds up the tongue and aids in holding back a vehicle on a downhill slope.
4. stout timbers held together by rods, bolts or clamps round the shuttering for a column to secure the parts together during the process of pouring and setting.

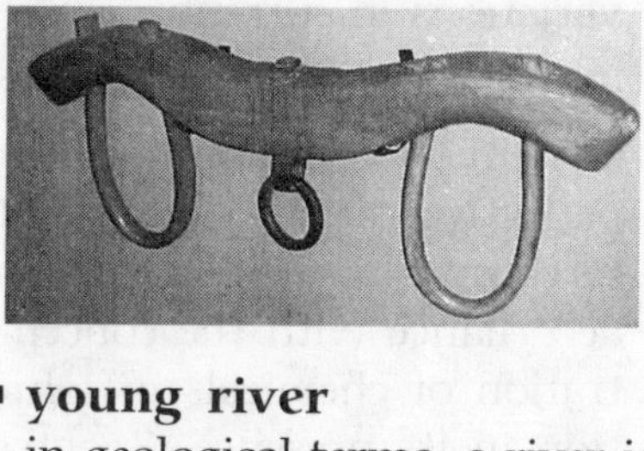

- **young river**
in geological terms, a river is young when it is actively eroding its channel.

z
the natural logarithm of the ratio of the two estimated standard deviations.

zaid
the season between Rabi and Kharif.

zawafa
an Egyptian water lifting device used for lifts exceeding two metres. It is a modern version of the chain and bucket type of Persian wheel but its design results in the water being lifted only a small distance above the delivery level.

zeeman effect
the splitting of spectral line by a strong magnetic field.

zener diode
a type of diode that is designed to conduct in reverse direction when its breakdown voltage is reached, maintaining a relatively constant voltage despite variations in current.

zeolite
a chemical compound so imperfectly bound together that its composition will change in accordance with the concentration of chemicals in solution in its presence. Zeolites are used in water softening processes.

zeolite method
a system for removing hardness (magnesium and calcium) from water. It consists of a tank containing a bed of quartz gravel. The zeolite takes up the magnesium and calcium and gives off sodium, which combines with the sulphate and bicarbonate, separated from the magnesium and calcium, to form harmless baking soda.

zeolites
a group of light coloured, hydrous silicates containing potash and soda; soluble and could play an important role in the base exchange in soils.

zero air voids unit weight
the weight of solids per unit volume of a saturated soil mass.

zero discharge layout
design of an industrial plant to release no effluent, whether cooling water (by closed recirculation system) or chemical effluent.

zero drift
movement of the zero point of an instrument in one direction, away from the correct zero, causing errors in all readings, although the differ-

ence between any two readings may be correct.

- **zero growth**
a state of the economy in which there is no overall growth in the level of economic activity.

- **zero index moisture**
that index of moisture when the precipitation is just adequate to supply all the water that would be needed for maximum evaporation and transpiration in the course of a year.

- **zero moisture index**
the index of moisture when the precipitation, analysed month by month, is just adequate to supply all the water that would be needed for maximum evaporation and transpiration in the course of a year.

- **zero plane displacement**
an empirically determined constant introduced into the logarithmic velocity profile to extend its applicability to very rough surfaces or to take into account the displacement of a profile above a dense crop.

- **zero point**
on a scale, the point separating positive scale values from negative scale values; may be arbitrary, that is, assigned according to convenience, or absolute, that is, having unique significance in reality.

- **zero tillage (no-till)**
the extreme in conservation tillage is no-till (zero tillage) where in the new crop is planted in the residue of the previous crop without any prior soil tillage or seed bed preparation and is usually possible when all weeds are controlled by the use of herbicides.

- **zero turgor**
turgor potential of zero, i.e., when the leaf water potential just equals the osmotic potential of a cell.

- **zerophytes,**
pants which can subsist with a small amount of moisture, as a desert plant or plant under conditions where excess of salts makes it difficult to take in water.

- **zeta potential**
the difference in potential between the fixed layer of solvent held to the surface of the solid in a hydrophilic gel, and the freely mobile layer which extends into the bulk of the solvent.

- **zg**
symbol for the degree of decomposition of humus. The ratio of the humic acid content to the loss on ignition, in per cent.

- **zig-zag separator**
an air column separator that is either zig-zag in shape or has baffles on alternate sides of the vertical shaft.

- **zig-zag system**
system in which field drains as well as mains are constructed zig-zag to reduce high velocities.

- **zingg bench terrace**
type of bench terrace suited for moisture conservation in dryland agriculture. It has an earthen embankment and a part of the terrace interval immediately above the embankment is bench levelled. Run-off from the sloping area is collected and absorbed on the levelled area.

- **zo**
a sub script symbol for a zoogenic horizon.

- **zonal flow**
the movement of air along the parallels of latitude.

- **zonal soils**
are the dominant soils of the climatic zones. Their nature is primarily determined by the climate of the zone and consequently by the vegetation of that climate. The influence of other factors of soil formation is either masked or is rendered subordinate to the dominating influence of climate and vegetation. These soils have well developed profile characteristics reflecting the influence of dominating climate and vegetation.

- **zonation**
an arrangement of different plant communities in more or less definite zone, following one another in a constant order due to a continuous change in the habitat factors.

- **zone**
1. in masonry dams, it is referred to a vertical portion between two horizontal lines of its cross-section assumed for stress analysis.
2. in earth dams, different material types are placed in zones.

3. in geology, used in the same sense as horizon to indicate a certain geological level or chronological position, without reference to the local attitude or dip of the rock.

- **zone grouting**
  a process of construction of a grout curtain whereby a whole section of the curtain is stage drilled and grouted to a limited depth before any holes are drilled and grouted beyond that depth.

- **zone of aeration**
  that portion of the lithosphere in which the fractional interstices of permeable rock or earth are not (except temporarily) filled with water under hydrostatic pressure; that is the interstices either are not filled with water or are filled or partly filled with water that is held by capillarity. The zone of aeration comprises the area where fringe water, intermediate water and ground water may be found.

- **zone of capillarity**
  an area that overlies the zone of saturation and contains capillary voids, some or all of which are filled with water that is held above the zone of saturation by molecular attraction acting against gravity.

- **zone of intermittent saturation**
  a layer of soil that lies just below the surface layer of soil and which will contain ground water after rainfall, but which will tend to dry out during periods of drought. Thus, the zone contains water only intermittently and the frequency of rainfall will determine the periods during which the zone is saturated.

- **zone of rock flowage**
  that part of the lithosphere in which all rocks are under stresses exceeding their elastic limits. In it the rocks undergo permanent deformation and are said to flow, hence in this zone interstices are absent or insignificant.

- **zone of rock fracture**
  the upper part of the lithosphere, in which rocks are under stresses less than those required to close their interstices by deformation of the walls of the interstices. It is the part of the lithosphere between the surface of the earth and the lower limit of porous water bearing rock formations.

- **zone of rock-flowage-and-fracture**
that part of the lithosphere lying between the zone of rock fracture and the zone of rock flowage. In it the strongest rocks behave as in the zone of rock fracture and the weakest rocks as in the zone of rock flowage.

- **zone of saturation**
that part of the lithosphere in which the functional interstices of permeable rock or earth are completely filled with water under positive hydrostatic pressure, that is, pressure in excess of atmospheric pressure. It is the portion of the lithosphere that supplies water to springs and wells and which contains the ground water.

- **zone of weathering**
that depth, or thickness, of the surface crust of earth in which weathering has taken place; includes soil and parent material of soil.

- **zoning valves, or control valves**
valves installed in longer sprinkler laterals to permit the shutdown of small segments of the distribution system for repairs or alternations, without curtailing the delivery of water to the rest of the land.

- **zoogene**
a soil or horizon one of the primary factors in the formation of which is the action of animals, e.g., earthworms, arthropod, etc.

- **zymosis**
1. fermentation.
2. reactions induced by enzymes.

- **zymurgy**
a branch of applied chemistry dealing with fermentation.

- **zz**
symbol for the coefficient of decomposition of the litter.